ROUTLEDGE LIBRARY EDITIONS:
20TH CENTURY SCIENCE

Volume 9

SCIENCE ADVANCES

SCIENCE ADVANCES

J. B. S. HALDANE

Routledge
Taylor & Francis Group

LONDON AND NEW YORK

First published 1947 by Routledge

2 Park Square, Milton Park, Abingdon, Oxon OX14 4RN
711 Third Avenue, New York, NY 10017, USA

First issued in paperback 2016

Routledge is an imprint of the Taylor & Francis Group, an informa business

Notices
Practitioners and researchers must always rely on their own experience and
knowledge in evaluating and using any information, methods, compounds, or
experiments described herein. In using such information or methods they should
be mindful of their own safety and the safety of others, including parties for whom
they have a professional responsibility.

Product or corporate names may be trademarks or registered trademarks, and are
used only for identification and explanation without intent to infringe.

British Library Cataloguing in Publication Data
A catalogue record for this book is available from the British Library

ISBN-13: 978-1-138-01356-8 (hbk)
ISBN-13: 978-1-1389-8140-9 (pbk)
ISBN-13: 978-0-415-73519-3 (Set)
eISBN-13: 978-1-315-77941-6 (Set)
eISBN-13: 978-1-315-77931-7 (Volume 9)

SCIENCE ADVANCES

by

J. B. S. HALDANE

LONDON
GEORGE ALLEN AND UNWIN LTD

FIRST PUBLISHED IN 1947
SECOND IMPRESSION 1949

PRINTED IN GREAT BRITAIN
in 12-Point Fournier Type
BY UNWIN BROTHERS LIMITED
WOKING

PREFACE

THE majority of these essays have appeared in the *Daily Worker*, one in the *New Statesman and Nation*, one in *Science and Society*, while the final article was published in *Nature*. In so far as one theme runs through them, it is the growth of pure and applied science. I have described some old and new discoveries and inventions, and the way in which they are being, or could be, used for the benefit of humanity or otherwise.

I anticipate two criticisms. I have sometimes repeated the same statement in several articles. This was inevitable, since they were written over a period of over four years, and also because the same facts are of importance in different contexts. And I shall be told that I have dragged in Marxism like King Charles's head. This is again inevitable if the writer thinks, as I do, that Marxism is the application of scientific method to the widest field so far achieved by man. If Marxism were taken for granted, or even if its general principles were widely understood in this country, such emphasis would be unnecessary. But the facts which I describe fit into a general framework, and Marxism is the best account of this framework which I know. If other writers on science can fit them into a better framework, by all means let them do so. But they are not isolated from one another, or from ordinary life, and it is a mistake to present them as if they were so.

I must thank colleagues who have helped me with facts, and readers of the *Daily Worker* who have criticized the articles and suggested topics. For this book is definitely a social product rather than the efflorescence of my own mind.

October 1944

CONTENTS

3. *Human Physiology and Evolution*—contd.

4

Medicine

5

Hygiene

6

Inventions

7

Soviet Science and Nazi Science

8

Human Life and Death at High Pressures

1

SOME GREAT MEN

Newton

Isaac Newton was born on Christmas Day, 1642, in the first
year of the Great Rebellion. Many people regard him as the
greatest man whom England has produced. But we mostly know
very little about him. This is largely due to ridiculous propaganda
about his life.

We think of Wordsworth's lines

a mind for ever
Voyaging through strange seas of thought alone

or the story that a falling apple set him thinking about gravitation.
He is made out to be a thinker divorced from ordinary life. As
a matter of fact he was a very skilled craftsman.

As a boy we know that he interested himself in conjuring
tricks. As a man he made the first reflecting telescope, whose
most important part is a concave mirror, not a lens. For this
purpose he had to experiment with various alloys, and to grind
his own mirrors. To-day all large astronomical telescopes rely
on mirrors rather than lenses to concentrate the light from
distant stars.

As an experimenter he showed that white light could be split
up into colours in two different ways. A glass prism refracts blue
light, that is to say bends it out of its original path, more strongly
than red light. And a thin film of air between two layers of glass
reflects light of a colour whose wave-length is just equal to twice
the thickness of the gap, or some multiple of this length. He also
worked on static electrification.

If Newton had never done a sum in his life he would be
remembered as a great technician and a very great experimental
physicist. In addition he was one of the greatest mathematicians,

perhaps the greatest, who ever lived. At the age of twenty-two he invented what is now called the differential calculus, and at the age of twenty-three the integral calculus. These were independently invented by Leibniz in Germany within a few years. Both these branches of mathematics were essentially tools for his great project of producing a mechanical account of natural events which would allow of their exact prediction.

The principles governing the flight of cannon balls were being worked out at this time. Newton showed that the moon in its course round the earth, and the earth and other planets in their courses round the sun, obeyed the same laws. This was a very great mathematical achievement. He also discovered the laws governing the flow of heat and of some fluids.

He believed that he was describing the properties of real matter in real space and real time. Some modern physicists say that this is impossible, and that we cannot know the real nature of things, but only devise theories which fit our experience more or less exactly.

They point out that the work of the last forty years proves that matter, space, and time are not what Newton thought them, and that newer theories enable us to predict what will happen more accurately than do Newton's theories. This is true. They go on to say that this shows that one can never know what matter, space, and time really are.

Certainly we can never know all about them. Lenin thought that the properties of even a single electron, the smallest bit of matter that we know, are inexhaustible. But most scientists think it nonsense to say that because we cannot know all about matter, we do not know anything about it.

How deeply Newton penetrated into the nature of matter is shown by a simple fact. The moving stars do not obey Newton's laws exactly. But their largest deviations from these laws are about one three hundredth part of the errors of measurement made by astronomers in Newton's day. If he had merely been making theories to fit the available observations, the theories would have been disproved as soon as more accurate measurements were made. He went far beyond the observations, a long way to the truth about matter.

Newton's world picture fitted in extraordinarily well with the ideology of the rising bourgeoisie. The need for it arose out of practical problems of his day, especially those of navigation and artillery. He pictured a world consisting of particles, each moving in a way which could be predicted accurately once the forces on it were calculated. In the same way society has been supposed to consist of isolated individuals, each behaving according to economic and psychological laws which governed their conduct.

As a result of the operation of these laws the universe was thought to behave in an orderly way. And if economic laws were allowed free play, with everyone acting according to his or her self-interest, bourgeois economists thought that society would function as perfectly as the stars in their courses.

History has disproved this latter theory, and about the beginning of this century it was shown that in some important respects matter did not behave in a Newtonian manner. If it did so, to take only one example, solid bodies would collapse. But modern physicists build on Newton's foundations, as Marx built on those laid down by such economists as Petty, Smith, and Ricardo.

Newton was not only a scientist. Like most other Englishmen of his time he believed that the Bible was inspired. He tried to interpret the prophecies in the book of Daniel. His attempt is interesting because, unlike most writers on such topics, he was very cautious in his deductions.

He was also a politician. In 1687 King James II threatened the liberties of Cambridge University, as those of Imperial College, London, are now being threatened. Newton argued the case for the University, and was elected by it to Parliament in 1689, as a supporter of the Revolution by which Dutch William replaced James II. He later became Warden and then Master of the Mint. These were key posts, as there was no paper money in those days, so the Mint was as important as the great banks to-day. His knowledge of metallurgy was put in the service of the state.

In fact Newton took part in the progressive political movements of his day. Like most great men, he was an all-round man. The idea that scientists should shut themselves away from every-day life did not appeal to him. He saw that science arises from, and ministers to, social needs, and he acted accordingly.

Marx

Sixty years ago Karl Marx died in London. Every year since his death he has had a greater influence on world history—above all since Lenin put his theories into practice in 1917. To-day even those who most abhor Marxism have to admit that he was a much more important historical force than such contemporary political figures as Gladstone, Disraeli, or Queen Victoria, or philosophers such as Herbert Spencer, Cardinal Newman, or August Comte, who seemed to be great in their own time.

He can justly be compared with contemporaries like Faraday, Darwin, and Pasteur, who are still influencing our lives and thoughts, because their ideas were important not only for their own time, but for many generations to come. These men applied scientific method to new fields. So did Marx.

The two volumes of his *Selected Works*, just published by Lawrence & Wishart, give some idea of the ground which he covered, and form an excellent introduction to his thought.

There were great socialists before Marx. They saw what sort of organization of society was needed. But they had not studied history deeply enough to analyse the process of historical change, and state the conditions by which socialism could come into being, as Marx and Engels first stated them in the Communist Manifesto.

There were great economists before Marx. But they were mostly content with describing the economic structure of society as they found it. Marx did not merely do this. He traced its origin and showed how it was decaying before his eyes, while the embryo of a new society was growing up within it in the form of the workers' organizations.

Marx was also a great historian and philosopher. Of philosophers he wrote, "Other philosophers have interpreted the world. The point is to change it." And here he was in agreement with the method of science. Hundreds of philosophers had interpreted the motions of the stars and other moving bodies. Galileo started experimenting, that is to say changing the motion or rest of material objects.

He found that Aristotle's and St. Thomas Aquinas's theories did not work. Ever since then experiment has been the method which scientists applied wherever it was possible, the method which gives the most certain results.

Academic philosophers had tried to explain the world starting from our sensations, and some of them concluded that the world consisted of nothing but sensations. Marx saw that we are just as closely related to the world by labour which changes it, as by sensation, which only copies it, and that a philosophy in which labour is not as important as sensation is of little value.

In the same way we can only get to understand the nature of society by trying to change it. No living man has a clearer grasp of the nature of society than Stalin, who has played a leading part in two great changes, the overthrow of capitalism, and the building up of socialism. Marx learned the true nature of class society from his early revolutionary work.

Just as Darwin applied scientific method to the problem of man's ancestry, and Pasteur to that of his diseases, Marx applied it to history, politics, and economics. In each case the result of the analysis was at first sight humiliating.

It was pleasanter to believe that we were made in God's image than that we were descended from monkeys, to regard an epidemic as a punishment from God rather than a result of a faulty water supply. So it hurt human pride to be told that history was determined by economic causes rather than by the ideas of great men, the judgments of God, or the racial soul rooted in blood and soil.

But humility is a condition for progress. If we believe that our ancestors were monkeys we can hope that our descendants will surpass us beyond our wildest imagination; if we know the material causes of disease we can hope to abolish all diseases as we have abolished many. If our history, laws, and morality rest on an economic basis, we can see the way to a progress which still seems impossible to many people.

By studying the laws of change in their most general form, Marx and his friend and colleague Engels not only illuminated history, but science. They did this in two ways. In the first place scientific discoveries are part of the historical process, and depend on productive forces and relations.

Newton's work was possible because people needed exact knowledge of the movements of the stars for navigation, and of cannon balls for war. He showed that they obeyed the same laws. Darwin could make his great generalizations because the exploitation of colonies had disclosed the distribution of living animals and plants through the world, and the development of mining had disclosed the order in which fossil animals and plants had appeared and died out in the past.

In the second place, material systems develop and perish according to dialectical principles like those which hold for human institutions. Engels was almost alone in his time in thinking that chemical atoms were not indestructible. Rutherford showed that they are born; and that they are destroyed, not usually by external forces, but by their own internal stresses.

Marx did not live to see his theories applied by Lenin. Maxwell did not live to see Hertz, Lodge, Marconi, and others apply his theory of electromagnetic waves to radio-communication. Leninism is Marxism developed by the experience of socialism in action. But it is still Marxism.

Outside the Soviet Union there is a wide and growing distrust of science. The intellectual leaders of the capitalist world, both within the churches and outside them, tell us that science is leading to increasing unhappiness, of which wars are the worst but not the only symptom, because it is applied to machines and not to the regulation of human conduct.

They are right up to a point. Marx said much the same a century ago. But he took the decisive step of showing how scientific method could be applied to human affairs on the broadest scale. So far it has only been so applied in the Soviet Union.

We celebrate the anniversary of the great teacher who has shown us the way out of our present distresses, who has demonstrated that there are no limits to the application of science. We can best honour his memory by doing all that we can to hasten the day when Marxism will be the guiding principle in the government of the country in which Marx spent most of his immensely fruitful life.

Archimedes

The Eighth Army has taken Syracuse, a city which, when it was free, broke the imperialism of Athens, and held up that of Rome for many years. It is therefore fitting to commemorate the greatest citizen of Syracuse, and probably the greatest Sicilian.

Archimedes was killed in 212 B.C., at the age of seventy or over. He was not only a first-rate mathematician, but a first-rate scientist. As a mathematician, he broke away from the formal and rather narrow methods of Greek geometry, and invented something very like the integral calculus.

Among the propositions which he proved was that the area of a sphere is the same as that of the curved part of a cylinder which just fits round it. The design of a sphere in a cylinder was engraved on his tomb. He also showed that π, the ratio of the circumference of a circle to its diameter, is between $3\frac{1}{7}$ and $3\frac{10}{71}$.

But his most important work was the founding of the science of hydrostatics. According to the story usually told, King Hieron of Syracuse had bought a crown which was alleged to be of pure gold. The king suspected that it had a core of mere silver, but did not want to cut it open, so he asked Archimedes how to decide the question.

Archimedes hit on the answer in the public bath, and was so excited that he rushed through the streets to the palace without dressing. If this is true, he was the first absent-minded professor in history, and the crown was worth more than all other crowns put together.

He saw that a pound of silver occupies nearly twice as much space as a pound of gold, and that the bulk can be measured by putting each in a full vessel, and measuring how much water runs over. So he compared the bulk of the crown with those of an equal weight of gold and an equal weight of silver.

This led at once to the notion that every substance has a characteristic density, or specific gravity. Thus a cubic inch of gold weighs 19·25 times as much as a cubic inch of water, while a cubic inch of silver weighs 10·5 times as much. So a silver crown displaces nearly twice as much water as a golden crown of the same weight, and a mixed one displaces an intermediate amount.

For thousands of years before Archimedes merchants had been weighing and measuring. Weight and bulk are examples of what are called extensive properties of matter. They add up. Two pounds and three pounds make five pounds, and so on.

But density is an intensive property. It does not add up. If you mix a metal with a density of 6 and one with a density of 10, you do not get an alloy with a density of 16, but usually with one somewhere between 6 and 10.

Modern physics and engineering are based on intensive properties which can be measured. Here are a few of them. Temperature, electrical conductivity, heat conductivity, hardness, elasticity, refractive index, albedo (fraction of light reflected) melting-point, boiling-point, solubility in water, magnetic susceptibility, coefficient of thermal expansion, viscosity.

Any engineer could mention dozens more. There are other extensive properties besides weight and bulk, such as heat content, entropy, electrostatic capacity, and so on. But physics could not start without the measurement of the intensive properties.

Archimedes went on to found the science of hydrostatics, and did it so well that some of his propositions are taught to-day, almost without change. And he made a beginning with statics, introducing such fundamental ideas as that of the centre of gravity.

He laid the foundations of physics, but very little was built on them until Stevin continued his work on statics, and Galileo founded dynamics, in the sixteenth century A.D. Science died out when the free Greek cities such as Syracuse were conquered by the Romans, and was born again in the United Provinces of the Netherlands, and in the free cities of Italy.

We can see why it died out from the life of Archimedes himself. He invented a number of machines, including a screw for raising water, and others used in defending Syracuse against the Romans. He is said to have set fire to their ships with a concave mirror, but this is no more possible than the heat ray in Wells's *The War of the Worlds*.

He refused to write accounts of any of these inventions, except a sphere for demonstrating the motions of the planets. He regarded them as beneath the dignity of a philosopher. This attitude to manual work was inevitable in a society based on slavery. But the gap between thought and manual work was not very wide in the Greek cities, where many citizens were craftsmen, and if they owned a slave or two, worked beside them at the bench.

As the Romans conquered the Mediterranean basin, and made millions of slaves, the gap became so great that science died out. Eighteen centuries later, in Holland and Italy, craftsmen once more became leading citizens, and science started again.

There was another reason why science decayed under the Roman Empire. Unemployment developed among the free population of Rome and other great towns, while the slaves were worked to death. The government paid for the building of temples, baths, circuses, and other buildings to give work.

The historian Suetonius tells us that an inventor approached the emperor Vespasian with a machine for moving heavy stones. He turned it down because it would have displaced labour.

Science only flourishes in special conditions. Italian science has been decaying for a generation or more, and Fermi, the greatest living Italian scientist, is a refugee in America. Sicily still produces mathematics, but little or no science.

If Archimedes's countrymen are allowed to decide their own destinies, the liberation of Sicily may make it possible for Sicilians once more to serve humanity as Archimedes served it.

Copernicus

The four-hundredth anniversary of the death of the great astronomer, Nicholas Koppernik, or Copernicus, fell in May 1943. Both the Polish and German governments are trying to claim him as a glory of their nationality. Actually his father had a Polish name, his mother a German. He was born in Torun or Thorn, a city belonging to the Hansa League of towns engaged

in trade, which was predominantly German. But the city government had recently accepted the protection of the Polish king
against the Teutonic Knights. The conflict between the rising and
still progressive bourgeoisie and the feudal knights had cut clean
across national divisions.

Koppernik probably learned Polish before he learned any
other language, so the Polish claim to him is a little stronger
than the German. But he spoke seven languages, wrote in Latin,
and probably thought in Latin, if only because the Catholic
Church, of which he was a canon, had the monopoly of education. I mention these facts because we cannot possibly understand
history if we think that the distinction between nationalities was
always as important as it is now. And we cannot talk sense about
the future if we suppose that it always will be.

Astronomy was naturally based on the idea that the sun and
stars went round the earth once a day. More accurately, if the
"fixed" stars go round once in a sidereal day, i.e. $366\frac{1}{4}$ times a
year, the sun lags one round per year, so there are $365\frac{1}{4}$ ordinary
or solar days per year. The moon lags still more, missing one
round each month. Mars, Jupiter, and Saturn lag less than the
sun, but Venus and Mercury never move very far from the sun,
sometimes appearing as morning, sometimes as evening stars.
On this basis you can calculate the positions of the sun and planets
pretty well, especially if you make further allowances according
to the theories of the Greek-Egyptian astronomer Ptolemy, who
lived in the second century A.D.

The churches still calculate Easter according to a scheme based
on Ptolemy's theory. That is why Easter and Whitsun were so
late in 1943. The full moon on whose appearance Easter is based
turned up a few minutes earlier than it should have done according
to Ptolemy. And this happened to make a difference of a month.
Unfortunately the churches are equally out of date regarding still
more important matters.

Copernicus showed that the directions of the sun, moon, and
stars could be calculated if the distant stars were fixed, and the
earth and other planets went round the sun in circles, while the
moon went round the earth, and the earth spun on its axis once
a day. This theory has had to be improved in detail. The distant

stars move, but they are so far off that the apparent shape of the constellations hardly changes in a thousand years. Kepler showed that the orbits of the earth, moon, and planets were roughly ellipses, not circles. Newton showed that the elliptical paths were due to gravitation. On his theory they would be exact ellipses but for the pulls of the planets on one another. The effects of these pulls were calculated and checked by observation. Thus when Jupiter and Mars are in a line with the earth, Jupiter pulls Mars further away than it should be on Kepler's theory. Finally, Einstein has made further corrections, and doubtless there are more to come, for the moon may be several seconds late or early as compared with the best calculations so far made.

Copernicus, or more probably a friend who added a preface to his book, wrote that the stars moved as if his theory were true, but did not claim that it was true. This may have been done to avoid being prosecuted for heresy like Galileo. Or the writer may have been a positivist like Mach and other philosophers attacked by Lenin in *Materialism and Empiriocriticism*.

According to positivists we have certain sensations, and it is a matter of convenience what theory we make to explain them. If two theories work, the simplest is the best, and it is silly to argue which is right. If this were true, it would make no difference to us whether the earth was really fixed or spinning on its axis, except in so far as it simplified astronomers' sums. Actually it makes a lot.

If a north wind blows from the pole, how will it move an ice floe? If the earth is fixed, the ice will move due south. If it is turning, the ice will not have enough eastward velocity to keep up with the sea as it moves south. It will be left behind, like a man running outward from the centre of a revolving disc, and will drift south-west, not south. This theory was verified by Papanin's polar expedition. It had been proved long before for air and water by a study of winds and ocean currents. But as wind has a bigger effect on ice than water, since, in order to make a current, the water must be set in motion down to a fair depth, the Soviet expedition was able to make particularly accurate observations in support of this principle.

Again, if the earth is still, a free and perfectly balanced gyroscope set to point upwards should still do so an hour hence.

If the earth spins, it should always point in the same direction in space, i.e. at the same "fixed" star. Actually it nearly does the latter, but not quite exactly, as some friction is unavoidable. But if the earth did not spin, the use of gyrostats for controlling aeroplanes would be a good deal simpler. A whole number of other phenomena measurable on earth show that its rotation is a reality, not a mathematical convenience. If we lived underground, and had never seen the sun or stars, but the sciences other than astronomy had developed as they actually did, the earth's rotation would probably have been discovered in the last century, but it would still be regarded as probable rather than proved.

Copernicus did not prove it. He made it so probable that he let loose a landslide of proofs. And he played a great part in showing man his real place in the universe, which consists of matter such as we know on earth, behaving in ways that we can understand and predict.

Landsteiner

Karl Landsteiner has just died in New York at the age of seventy-five, working up to the last moment. He was one of the very greatest scientists of our time, being a first-rate chemist as well as a first-rate biologist.

Almost all his work was concerned, in one way or another, with what is called humoral immunity. Every animal and every plant makes proteins according to its own special patterns. If a protein of a foreign pattern is injected into an animal, the receiver usually forms what are called antibodies, which can be found in its blood. These combine with the foreign protein if more is injected. The results may be good or bad. For example, after several injections of snake venom or diphtheria toxin a man generally has enough antibodies to allow him to handle a snake of that particular species without danger if he is bitten, or to expose himself to infection with diphtheria without danger of serious illness. But immunization may be dangerous. If a man has had an injection of horse serum from an immunized horse as a preventive against tetanus, he develops antibodies against the proteins of horse serum, and a second injection may make him ill, whereas the first did not. He reacts to the injection of a previously harmless substance with fever and other symptoms like those which he puts up in fighting an infection.

Landsteiner brought these previously rather mysterious phenomena into the sphere of chemistry. He altered proteins by attaching dyes and other well-defined chemical groupings to their molecules, and investigated how far the antibodies developed were specific. The antibody fits the antigen, as the foreign protein is called, much as a key fits a lock. However, some keys will fit a number of fairly similar locks. The specificity is not absolute. These chemical methods have been improved and extended. We now know that the combination of antigen and antibody follows chemical laws, and last year Pauling, a Californian chemist, was able to make artificial antibodies in his laboratory without using a living man or animal to make them. They were not very efficient, but probably in another generation or so most antibodies will be made in this way, which may remove some of the prejudice against them.

Landsteiner was more concerned with the harmful than the useful effects of humoral immunity. He showed that the injection of blood from one human being into another was generally dangerous because the blood of the recipient contained antibodies which damaged the corpuscles of the donor. He found that human beings belonged to different groups whose corpuscles carry one, both, or neither, of two antigens called A and B, and that transfusion between members of the same groups was almost always safe, while blood from one, but only one, of the groups could be safely injected into members of all the others. At the same time that Landsteiner made this discovery in Vienna, Jannsky obtained similar results in Prague. Landsteiner and his colleagues next showed that membership of a blood group was inherited according to simple laws. Actually he made a slight mistake, which was corrected by Bernstein of Göttingen, who, like Landsteiner, was of Jewish origin and escaped to America. Landsteiner was big enough to accept Bernstein's work without clinging to his old half-truths as lesser men have often done.

Hirzfeld and his wife, Polish doctors working with the allies in Salonika in 1917, found that members of the four groups were found in all the peoples studied by them, but in different proportions. Thus the Nazi doctrine that the blood is something peculiar to each race was disproved while Hitler was still a corporal. You may be a blue-eyed blond "Aryan," but a blood transfusion from another such will kill you, while a pint from a negro will save your life.

Landsteiner went on to study other differences of the same kind between different human bloods. Along with Levine he discovered two other antigens called M and N which are of no importance in normal transfusions, though of great interest to students of heredity. Some people thought that his work had become of merely academic interest. But in his last years he was one of a group in New York which made a discovery that may save even more lives than that of the blood groups. The blood corpuscles of the Rhesus monkey carry an antigen called Rh, and if they are injected into guinea-pigs the latter develop antibodies which destroy them.

Landsteiner found that about 84 per cent. of Europeans and Americans of European stock have the Rh antigen in their blood corpuscles, but 16 per cent. do not. On the other hand, all or almost all Chinese and Red Indians possess Rh.

Now if blood is transfused from a person with Rh to one without it, no harm is done on the first occasion, but the recipient sometimes (fortunately not always) develops antibodies against it. If so a second injection from a person with Rh is dangerous and often fatal. The red corpuscles are treated as foreign bodies and rapidly destroyed. Their debris may kill him by clogging the kidneys, or in other ways.

This accounts for a good many of the hitherto unexplained deaths following blood transfusions. But there is an even worse danger. If a mother without Rh marries a man with it, most of the children usually inherit Rh from the father. In a minority of pregnancies Rh passes from the unborn baby into the mother, and she develops antibodies against it.

If she loses blood during childbirth and has a transfusion from her husband or anyone else with Rh, this may kill her. Further, the antibodies in her blood may get into that of the baby and destroy its blood corpuscles. It may die before birth, or be born with jaundice. Or it may develop jaundice within a few days of birth; and if so it generally dies.

Such babies, if born alive, can often be saved by a transfusion of blood from a suitable donor. There is at present no way of saving them before birth, though an obvious method has been suggested. But this can only be achieved after several years of hard laboratory work and a year or so of experimental treatment. So Landsteiner's death probably involves the death of thousands of babies.

It has been estimated that, including still-births and abortions, one baby in four hundred dies from this cause. Dr. Mollison, of the Sutton Blood Storage Centre, has published a number of pathetic cases where every child of two healthy parents died from it before or soon after birth. But we have good hope that this sort of thing will be prevented in future.

Landsteiner's work was recognized all over the world. I had the honour of presenting his claims when he was elected a foreign member of the Royal Society. He was born an Austrian and died an American. But he was a servant of all mankind, saved the lives of thousands who had never heard of him, and will save the lives of thousands still unborn.

NOTE.—Since this article was written, it has been shown that there are several slightly different Rh substances, and that a marriage between persons with different ones may lead to the death of some or all babies. However, the above account is correct as far as it goes, being the truth, but not the whole truth.

de Geer

Baron de Geer, one of the world's most original scientists, has recently died in Sweden. His life work was to provide fairly accurate dates for things that happened before men kept records.

The earliest date which we know accurately is 2283 B.C. On March 8th of that year the sun was totally eclipsed near the city of Ur in Iraq, Abraham's home town according to the Bible; and shortly afterwards the city was taken by the Elamites from what is now south-western Persia. The date of the eclipse can be calculated, for the movements of the earth round the sun and the moon round the earth are very regular. This date agrees fairly well with that calculated from lists of kings, and these may be roughly correct for another thousand years or so back; but even in Iraq and Egypt, history peters out into legend somewhere about 4500 B.C. which is the rough date of the great flood that certainly swamped Iraq, and equally certainly did not swamp the whole world, as claimed in the story of Noah.

De Geer derived his dates, not from stone monuments or clay tablets, but from the soil of his country. Much of southern Sweden is covered with clay, and a cutting shows that this consists of horizontal layers. By comparison with the Alps it is clear that this clay was laid down in fresh water near the edge of a melting ice sheet, and each layer was laid down in the spring of a different year. In a warm year a lot of ice melted and a lot of mud was deposited; in a cold year the layer of clay was thin. At any particular place in southern Sweden there may be only a dozen or so layers. But by comparing hundreds of soil sections de Geer was able to get a continuous series up to Lake Ragunda in central Scandinavia, which was drained in the eighteenth century. There is some uncertainty about the dates of the recent layers, and this will not be settled till another mountain lake is drained. But there is no doubt that somewhere about 8000 B.C. the edge of the ice was at the present site of Stockholm, and that after this it retreated fairly quickly into the mountains. In 10000 B.C. the ice still covered the southern tip of Sweden, and about 18000 B.C. it stretched across the Baltic to northern Germany, where its southern edge left huge moraines.

Naturally no human remains have been found in these deposits. But we know that in central and western Europe and Britain the period of increasing warmth when the ice was rapidly retreating was the mesolithic period when men were changing their technique of making stone tools from chipping to polishing. During the last ice age the inhabitants of France hunted reindeer with implements of chipped flint, and drew pictures of them on bones.

As the climate improved they began to polish their stone tools, and to keep flocks and grow crops. In fact fifteen thousand years ago the people of Western Europe were in the same stage of culture as the most primitive savages alive to-day.

De Geer's pupil Antevs and others have made similar records of the mud layers in Canada and the northern part of the United States. The dating is not so certain, but the time scale is roughly the same. A similar method of dating by tree rings has been worked out by Douglass and others for Arizona and Persia. In a fairly dry country with a variable rainfall a tree makes a great deal of wood in a rainy year, and much less in a dry one. Where trees live for several centuries they give a good record of climate, and an expert can easily date a log showing a hundred or so annual layers of hard and soft wood. This time scale so far only goes back for about 3000 years. But it can be used to date "Indian" houses in Arizona quite accurately, and hence tools found in them, although their makers left no written records, and their spoken tradition is of miracles rather than history, and has no dates.

A knowledge of these facts is important for the following reason. We are apt to take it for granted that the kind of society in which we live, and which goes back for some thousands of years beyond our historical records, is natural for man. To many people class divisions and the ownership of property other than clothes and tools seem normal. They are not normal, but very recent innovations. Men, that is to say, creatures like ourselves, using tools and fire, have existed for at least half a million years, judging from the depths of mud laid down by water and ice, and from other evidence. During almost all this time men have been hunters and food gatherers living in small bands.

The great changes which began ten thousand years ago are still gathering speed. We have no idea of what men will be like ten thousand years hence, except that they will be unlike ourselves, and probably just as unlike the beings prophesied by Wells, Shaw, or Stapledon. But Marxists have solid and scientific grounds for believing that the next stage of human development will be worldwide and peaceful socialism.

Eddington

The evil that men do lives after them,
The good is oft interred with their bones

said Mark Antony at Caesar's funeral; and after listening to the
B.B.C. and reading several newspapers on the late Sir Arthur
Eddington, I am almost inclined to believe it. If I had not known
Eddington and followed his work, I might believe, from these
accounts, that he had done little more than comment on Einstein's
theory, and write a certain amount of idealistic philosophy. It is
not for either of these things that posterity will remember him, as
I think it will.

He was an astronomer by profession, and for thirty years
directed the Cambridge Observatory, where a large amount of
very accurate work is done. That meant that he had to be a
skilled manual worker, to be able to deal with apparatus which
is as much more delicate than an ordinary camera as a camera is
more delicate than a carving knife.

He did not merely explain and expand Einstein's theory. He
checked up on it. Einstein said that light had weight, and was
therefore bent out of its path when passing near to a heavy
object. The only object which is heavy enough to bend it to a
measurable extent is the sun, and light from a star passing near
to the sun can only be photographed during a total eclipse. In
1919 Eddington took out an expedition to the island of Principe
off the coast of West Africa, and photographed the stars which
were visible near the sun when it was eclipsed. Some months later
the same stars were photographed at night with the same camera.
In the night photograph their images were further out from the
centre of the plate, showing that the sun had bent the rays inwards. He found the effect predicted by Einstein, which is just
twice that which might be predicted on a basis of common sense.
The measurements were so accurate that no better ones have been
made in the ensuing twenty-five years. As a result physicists now
think that light is much more like ordinary matter—more
material if you prefer that expression—than they did before.

He also did a large amount of statistical work on the movements of the so-called fixed stars relative to the sun, based on

thousands of observations by himself and others. He made a number of calculations on the internal constitution of the sun and other stars, which later workers have modified, though they accept most of his principles.

His greatest contribution to astronomy was probably his discovery that the luminosity of a star depends mainly on its mass, a most surprising fact which however agrees well with his theories. One can only determine the luminosity of a star when one knows its distance. For example Arcturus gives out four times as much light per minute as Sirius. But it is nearly five times as far away, so it only appears to us about a fifth as bright. It is hard to determine the distance of a star, but still harder to determine its mass. This can only be done accurately when a pair of stars are revolving like a pair of dancers round their common centre of gravity. But very careful observations are needed to get the mass correct even within ten per cent.

Much of Eddington's writing was idealistic. He believed, with Kant, that one can know nothing about the real nature of matter, and that almost or quite everything which we think we know about it is really the product of our own minds. I think he was wrong, but to those Marxists who would condemn him root and branch I should like to repeat Lenin's words: "Philosophical idealism is nonsense only from the standpoint of a crude, simple, and metaphysical materialism." Eddington was so impressed by the rational character of reality that he sometimes forgot that it had any other characters. Nevertheless I think he was far nearer the truth than a materialist who takes the various different characters of matter for granted, and refuses to believe that they have any rational connection. Eddington found rational connections between the properties of stars and of atoms. He was profoundly impressed by the fact that the real is the rational, and did not realize that one can believe this without being an idealist. Like many great scientists, he was a Quaker, and this certainly influenced his thought in an idealistic direction.

The *Daily Worker* obituary notice stated that "he popularized the idealist view that the behaviour of matter is uncertain and unpredictable, a view which Marxists hold to be wholly unwarranted and due to misinterpretation of the facts." If that is true

I am not a Marxist. Nor was Lenin when he wrote in *Materialism and Empiriocriticism* that "the sole property of matter with the recognition of which Marxism is vitally concerned is the property of being objective reality, of existing outside our cognition."

Imagine a blindfold man who is trying to find out just where an electric bell is. He can listen for it, but is only allowed to feel for it with a revolving brush such as a chimney-sweep uses. He will not be able to state exactly where it is, or of what shape. If he makes a lot of attempts he will be able to state its probable position, but no more. If he is an idealist he will say this proves that there is nothing really there. We are all in a similar position. We can locate things with light, but its wavelength limits our accuracy. We can measure them with a gauge, but its atoms are moving, and this again limits our accuracy. Fortunately we can measure distances of less than a millionth of an inch, so for most practical purposes we can reach certainty. But we cannot, for example, examine the inside of a particular molecule in a flame, and say that it will emit some light in the next millionth of a second. Its behaviour is uncertain and unpredictable. However, the behaviour of a thousand million such molecules is certain and predictable. As Engels put it, "One knows that what is maintained to be necessary is composed of sheer accidents."

It seems to me that we have to steer our way between two mistakes. Eddington's mistake was to think that because the behaviour of matter in small samples was uncertain and unpredictable, it was not real, or had free will. Our correspondent's mistake was to think that because a thing is real we can necessarily get complete information about it. I once asked Lord Keynes what he thought of the Swedish economist Cassel. "A very able man," he replied, "but he makes mistakes." Perhaps Lord Keynes and the Pope are infallible, perhaps not. I certainly make mistakes, and so I think did Eddington.

Unfortunately he was honoured for his mistakes by readers who could not follow his real work. But he was a great astronomer and a great physicist, and his positive achievements in these sciences will be remembered centuries hence, when any mistakes which he may have made are taken no more seriously than are Faraday's Sandemanian theology or Newton's interpretation of the book of Daniel.

Wilson and Bragg

When microscopes were invented some people hoped that by improving their lenses men would be able to see even the smallest objects. We now know that we cannot do this, because light consists of waves, and we cannot see things much smaller than the length of these waves, even with the best microscopes. Now atoms are very much smaller than light waves. About a thousand atoms in a row would be needed to make up one wavelength. So many scientists thought that atoms would never be seen or photographed, and a few thought that they were not real, but just convenient fictions useful to chemists.

The first man who can be said to have seen atoms is Professor C. T. R. Wilson, of Cambridge, in England. All people who live in mountainous regions have watched a cloud form as moist air moves up a mountain side. This is because the air expands as it rises, since there is less air above it to compress it. In expanding it does work and loses heat. So the water vapour in it condenses into drops of mist. But drops can only form on dust, or on atoms or molecules of gas with an electric charge. Wilson made a chamber in which dust-free air was made to expand suddenly, and he could see through a window the drops of mist which formed on the electrically charged particles. Now even a single atom, if it moves fast enough, will give electric charges to some of the molecules of the gas through which it passes, much as a glass rod can be charged by rubbing. The moving atoms shot out from radium leave tracks of mist which Wilson could see or photograph, and thus measure how fast they were going. Of course he only saw the atom in the sense that we can see a rocket at night by the track of sparks which it leaves in the air, or as Londoners in 1940 saw the tracks of cloud left in the sky by the aeroplanes which were fighting so high above their heads that they could not be seen directly.

If atoms were to be photographed when at rest, it was necessary to use something like light, but with less than a thousandth of its wavelength. The first man to do this was Sir William Bragg, who has just died. This great physicist went out from England to

Australia, where he worked for many years before coming back to England. In 1912 a German physicist, Laue, showed that when X-rays passed through a crystal they made a peculiar pattern on a photographic plate, and saw that this must mean that X-rays were trains of waves like ordinary light, but shorter. X-rays had long been used by surgeons for seeing through the living body, but no one knew what they were. Laue could not measure the wavelength, and no one can get very far in physics without measurement.

In the same year Bragg and his son showed that X-rays can be reflected by crystals, though they are not reflected by ordinary polished surfaces. But they are only reflected from crystals if they strike at certain angles. The crystals consist of layers of atoms arranged in regular patterns. If a beam of X-rays strikes the crystal so that a wave which has penetrated to the second layer and been partly reflected from it has travelled by a path longer by just one wavelength than a wave which has been reflected from the outer layer, and similarly for those which have penetrated deeper, the waves will be reflected. Just the same principle holds when light is reflected from a pearl, which is built in layers a thousand times thicker than those of most crystals. The different colours are reflected at different angles.

The arrangement of the atoms in a few simple crystals was known already, so this discovery enabled Bragg to classify X-rays by their wavelength, as different coloured lights are classified. This was of great value in medicine and surgery. The X-rays with long waves do not penetrate far into matter. So they are of little use for making photographs of broken bones or bullets in the flesh. They are stopped even by the skin, and their chief use is for treating skin diseases. Those of medium wavelength are used by surgeons. Those of very short wavelength will even pass through the human bones, but they can be used to photograph heavy metal objects such as the crankshafts of motor-cars or tanks, to detect hidden flaws.

Still more important was Bragg's work on the arrangement of the atoms in crystals. For nineteen years he was in charge of the laboratory of the Royal Institution in London. Many great men had held the post before him. It was there that Davy discovered sodium and Faraday electro-magnetic induction, on which the whole electrical industry is based. There too Dewar liquefied air and invented the "thermos" vacuum flask. Bragg and his pupils photographed many thousands of crystals with X-rays, and published diagrams and models. Thanks to them, mineralogists can now learn a series of facts as rational and orderly as Hindustani grammar instead of a series as irrational as the spelling of English words or the conjugation of Spanish verbs.

A bar of metal such as iron consists of a great number of very small crystals arranged in an irregular way, and often distorted by working. Many properties of metals have only been understood since the structure of these crystals has been determined. New alloys have been invented as a result of the information gained from X-ray photographs, and most great metallurgical firms employ an X-ray crystallographer.

Matter may be less regularly arranged than in a crystal, yet still regularly enough to reflect X-rays according to the laws discovered by Bragg. Here his pupils have carried on his work. Astbury in Leeds discovered how the atoms are arranged in wool and such tissues as human hair and nails. Bernal in London was able to photograph viruses, which cause some of the worst diseases of men, animals and plants, including smallpox and influenza, and are far smaller than bacteria, so that they cannot be seen with the most powerful microscope.

Some of the discoveries about the structure of matter which were made by this method merely confirmed what the chemists had long believed. For example the photographs showed quite clearly that the six carbon atoms in a benzene ring were arranged at the corners of a regular hexagon, and those of paraffins in a chain. But in a small number of cases the X-ray workers could not agree with the chemists. In these cases the chemists have always proved to have been wrong when they reinvestigated the matter. X-rays are now being used to investigate the structure of substances which have so far proved too difficult for the chemists alone.

Naturally Bragg's work was sometimes controversial. Shortly before his death he was in disagreement with the eminent Indian physicist, Sir V. Raman, on the interpretation of certain reflexions which were not predicted by his original theory. Bragg attributed them to imperfections in crystal structure, Raman to atomic vibrations. Posterity will decide which, if either, was right. Science progresses by such arguments as this, which generally suggest new experiments.

Another very important line of research is to see how the structure of a metal crystal is deformed when the metal is subjected to compression or tension. The change in shape and size of a whole block of metal may be due to the crystals slipping over one another, or to their changing their internal arrangement. We now know that both can occur, and this knowledge enables us to predict what particular composition and heat treatment will best enable metals to stand any given kind of stress. For the first time scientists can investigate what is going on inside an opaque solid body. This is as important for engineering as was the original discovery of X-rays for medicine, which allowed doctors to see the heart, lungs, bones, and other organs in a living man.

Sir William Bragg worked on many other subjects besides X-rays. He was an expert on sound, and in the war of 1914–18 he was one of those who designed the "Asdic" apparatus by which ships can hear U-boats at a great distance, and which played a great part in defeating the U-boat campaign, thus both saving Britain from starvation and enabling British armies to be sent overseas. He was also an extremely good popular lecturer, and disagreed strongly with those who think that the education of ordinary people in science is unimportant.

He was honoured both in Britain and elsewhere. He was elected President of the Royal Society, received the British Order of Merit, the Swedish Nobel prize, and many other foreign distinctions. His son, who helped him in his early work, and is continuing it, succeeded Rutherford as Professor of Physics at Cambridge. During his last year he played an important part in a conference on the application of science during and after the war. He was particularly fortunate in that his own work not only gave us new knowledge about the solid state of matter, but was immediately applicable to improving the quality of metals, textiles, and many other manufactured articles.

Boys

Among the British scientists who died in 1944 was C. V. Boys. My readers may not have heard of him, but in a sense they came in touch with him whenever thay paid a gas bill. For he designed the apparatus which determines the price of a cubic foot of gas.

Some scientists, including myself, are genuinely interested in theory. We see the implications of a theory, and are able to devise relatively simple experiments to check its results. For example, I predicted that the volume of air which I breathed would go down if I ate a couple of ounces of bicarbonate of soda. The drop was so large that there was no point in measuring the volume accurately even to 1 per cent. Others start out from action, and build up their theories round it. Boys was once asked why he did not employ a skilled mechanic to help him in constructing apparatus. He replied that his ideas only got into shape as the constructional work proceeded, and that this work helped the thinking process, and he would not get on any quicker by having it done for him. In the laboratory, at least, he was a better Marxist than I.

His most important piece of work was to weigh the earth, or in other words to determine the constant of gravitation. Here is the problem. We can measure the force which the earth exerts on a small metal ball. This force is the ball's weight. We have to compare this with the force exerted on it by another metal ball an inch or so away. This latter force is about a thousand-millionth of the weight, and Boys measured it correctly to about one part in three thousand.

To do this he designed an apparatus which was on view in the Science Museum at South Kensington before the war. The force is not measured directly, but two gold balls of a quarter of an inch in diameter are attached to a metal beam about an inch apart, and this beam is hung in a tube. Outside the tube are two lead balls of $4\frac{1}{2}$ inches diameter whose position can be altered, and the effect on the swing of the gold balls measured. The gold balls are suspended from a thread of quartz. Boys found that melted quartz could be drawn out into fibres, so fine that they could only be seen with a microscope, and stronger than steel wires of the same thickness would be if they could be made. To draw a thread while the quartz was still molten he used a cross-bow, also of his own design.

Incidentally he was the first to show that a quartz vessel could be made red-hot and quenched in water without cracking it; so a considerable industry arose directly out of his work. He finally showed that when the gravitational pull of the lead balls was re-versed, he got a deflection of $1\frac{1}{2}$ degrees in the average direction of the beam holding the two gold balls. Thus he determined the mass of the earth, and indirectly of the sun and other heavenly bodies. Every calculation of the mass of a star ultimately rests on Boys's work, carried out in a cellar at night to avoid shaking by traffic.

He also designed an apparatus to photograph lightning flashes in such a way as to determine their speed. As a flash is over in about one ten-thousandth of a second, a cine-camera is useless. Boys's camera has two rapidly moving lenses which take photo-graphs on the same plate. The images differ slightly, and enable the speed of the flash to be determined. Using a modification of this camera, Schonland, Malan and Collens in South Africa found that a lightning flash is far from simple. The first process is a "leader" flash going in steps of about fifty yards. Then follows a much brighter flash along the same track. The process may be repeated several times.

Boys never became a professor. He was an assistant professor at the Imperial College till 1897, and then became a Gas Referee. He was concerned with measuring the heating power of coal gas, while my father, who was one of his colleagues, measured its lighting power and the amounts of sulphur and other im-purities in it. The laboratory was always dirty. Both he and my father preferred to clean their own apparatus when necessary rather than trust the best charwoman in the world to dust it daily. The calorimeter measures the rise in temperature produced in a steady current of water by a steady current of burning gas. Now a cubic foot of gas contains less matter when hot than when cold. Boys allowed for this by automatically cutting down the water flow when the temperature of the gas rose. The apparatus is most ingenious and very simple. Boys always preferred mechanical to electrical principles, and anyone can understand his devices; but no one else had thought of them.

For over forty years Boys held this part-time job of Gas Referee, and made a very good income as an expert witness in cases involving patent law. His genius was very largely wasted, both from the point of view of pure science and from that of increasing the national wealth. His apparatus certainly secured a fair price for gas according to capitalist economics; for the cost per cubic foot depends on the heating power. But he would have done far more for the public good if he had designed more economical gas stoves, as he certainly could have done. And he would have done far more for pure science if he had devoted his skill to accurate measurements of the heat produced in well-defined chemical reactions.

There can be little doubt that he would have been far better appreciated in the Soviet Union, where the skilled workers constitute the nearest equivalent to an aristocracy that exists, than he was in Britain. For his qualities were essentially those of a craftsman. He would also have been better used. It makes a lot of difference to profits, but very little to production, whether one company or another can claim royalties for the use of a given patent. Where science is not planned "freedom" may mean in practice freedom to use one's talents on disputes about patents. Where science is planned, men of the requisite ability do work which solves fundamental problems of science, industry, or of both. Such men as Boys are rare. We should see that their talents are not wasted.

Milne

The progress of physics affects us most obviously by leading to new inventions such as electric lighting, radio, or refrigerators, or to very rapid progress in technique, for example the improvements in flying technique which render our lives rather precarious at the present moment. However, great theoretical developments not only lead to new techniques, but alter our outlook on the world, and thus affect history. Copernicus's new system of astronomy, according to which the earth is only one of several planets, and still more Newton's demonstration that the planets moved as a result of the familiar and measurable force of gravitation, certainly affected human thought. In particular they were necessary preliminaries to any attempt to apply scientific method to human history, as Marx did.

Very great advances in theoretical physics have been made during this century. Unfortunately they have been expounded to the public in this country in such a way as to make the universe appear more mysterious, instead of more intelligible. This is a sign of social and intellectual decadence. During periods when conditions are improving men feel that they can control their environments, and concentrate on expounding what is known about the world. A class which feels power slipping from its grasp consoles itself by thinking that the whole world is past understanding. One form taken by this intellectual defeatism is to say that the advance of science has proved that matter does not exist. This is correct if matter has the properties which we were taught at school. Its real properties turn out to be a good deal less simple from the point of view of a mathematician, but perhaps rather closer to those which our senses suggest.

Of course, very few working physicists are idealists. Matter seems real enough when you are working with it, though it may seem shadowy when you are doing sums about it. And some at least among mathematical physicists believe that they are making the universe easier to understand. One of these is Professor E. A. Milne, of Oxford. Most of his colleagues do not yet accept his views, but they are proving so fruitful that they may quite possibly be widely held, with minor modifications, a generation hence. Perhaps Milne's most important idea is that the universe is more rational than most scientists had thought, though not more so than Hegel and Engels believed.

The Greeks rationalized mathematics. The Egyptians knew that if they had a loop of rope divided into twelve equal parts by knots, and stretched it into a triangle whose sides were 3, 4, and 5 units long, the two short sides would make a right angle.

They took this as a fact, like many other facts which we cannot yet explain, for instance that all feathered animals lay eggs, or that there are higher mountains in Scotland than in England. The Greek Pythagoras showed that such a triangle must have a right angle because $3^2 + 4^2 = 5^2$, and that other right-angled triangles can be made with sides 5, 12, and 13; or 8, 15 and 17 units long, and so on. In consequence of such discoveries as these, mathematics has become a rational structure, not a mere list of rules.

Newton made a great step forward when he showed that the observed movements of the earth and planets could be explained if they and the sun attracted one another with a force varying inversely with the square of the distance. That is to say, the pull is reduced to one-quarter if the distance between two stars is doubled, to one-ninth if it is trebled, and so on. This law is not quite exact, but it is so near the truth that predictions from it came true until astronomers made observations with only one three-hundredth of the errors that they made in Newton's day.

Milne believes that he has proved that the inverse square law of gravitation must hold, except at very great distances indeed, as Pythagoras showed that the 3, 4, 5 triangle must have a right angle. Of course he does not start off from "pure thought," but from certain assumptions, of which the most important is that the universe would look much the same to an observer anywhere in it. That is to say, there is no end to matter, in the sense that an observer on some star would see other stars to the north of him, but only empty space to the south. Nor does matter thin out in any direction, though of course there are local aggregations of it, such as stars and nebulae. This is obviously a generalization of Copernicus's idea that our earth is not the centre of the universe.

The proof involves some fairly stiff mathematics, but nothing like so stiff as those of Einstein's general theory of relativity. Among the other consequences of Milne's theory are that the properties of space merely express inter-relations of matter and of light, and that one kind of geometry is more appropriate for light, and another for matter. As a result of the contradiction between these two sets of properties, the universe changes, though very slowly, so that the properties of matter are not the same now as they were, say, when the coal measures were formed, the rate of chemical changes having speeded up relative to that of certain physical changes.

Milne has been violently attacked for giving up the inductive method, going back to Aristotle, and so on. I think the history of science shows that there is a dialectical interplay between deduction and induction. In *The Marxist Philosophy and the Sciences* I gave a brief account of his theory from the Marxist angle, and he tells me that I have not misrepresented him. Since then he has developed his theory to cover electric and magnetic phenomena, and has succeeded in explaining facts which had formerly seemed quite unconnected.

Some readers will think that a working-class newspaper is a strange place to write about theoretical physics. If so, they forget that Engels did so in his not very ample spare time as secretary of the First International, and that Lenin did so after the failure of the Russian Revolution in 1905. Marxism can be applied to all branches of science as well as to economics and history, and no Marxist can neglect the progress of physics.

2

ANIMALS AND PLANTS

Newts

Boys and girls are coming home with newts in jam pots. I am going to write about them because they are among the few pets which can be kept without eating anything which human beings or even pigs could have eaten. A smallish earthworm every second day is the ration of each of my three newts, though they sometimes get an extra.

We have three species of newt in England. The common and palmate newts grow up to about 3 or 4 inches long, the crested newt to about 6 inches. The two small species spend most of the year on land, while the largest may not come out at all, though it generally does so in youth; but all go back to the water in spring as soon as they are old enough to breed, which is said to be at about three years.

On land their shape is roughly that of a lizard or crocodile, but in the water they have fins above and below their tails. The females' tail fins are merely efficient swimming organs, but the males' fins have a jagged edge like a cock's comb.

Even the common newt male is a gorgeous beast. Both sexes have a brown back and a bright orange belly. The male also has a parti-coloured tail. Its lower edge is bright scarlet with a broad strip of Cambridge blue above it. He uses it not only to swim but to dance in front of the female. He faces her, bends his tail through 180 degrees, and vibrates it like a pennant fluttering in the breeze, so as to direct a stream of water at her. So far the courtship is not very unlike that of a peacock, a pheasant, or many other birds where the male is particularly bright.

But what follows is much odder. In many water animals the male and female shed their spermatozoa and eggs into the water, and fertilization occurs outside the female. In some, and in most land animals, the male embraces the female, and fertilization occurs inside her. The newt does neither. When the male has excited the female by his dance, he leaves a fairly solid and elaborately patterned bundle of sperm, which she then picks up and thus fertilizes herself.

After this she lays an egg or two a day for a month or more. The eggs are surrounded by jelly, like those of frogs or toads. Each one is laid on the leaf of a water plant, and the mother then wraps the leaf round it with her mouth and back legs. The eggs hatch into tadpoles about ¼ inch long, with long feathery dark pink gills. The tadpoles eat water fleas and insect larvae, but if you want to keep them you must transfer the parents to another tank, as they will eat their children without hesitation. They also eat frog tadpoles.

Newts breathe through their skins under water, and must have plenty of pond weed to make the oxygen which they need. They also occasionally come to the surface for a breath of air, for they have lungs, and it is desirable to give them a bit of wood or bark to sit on.

About June the adults of the two small species lose their fins and spend more and more time on their rafts, if you keep them in an aquarium. When this happens they should be released in a damp spot not too far from water. For their skins, which in spring are thin and slippery, like human lips, become dry, and are no longer able to take up oxygen from the water. So they may drown in an aquarium. The young newts lose their gills in July or August, and should also be released. It is possible to keep them, but it is hard to obtain a supply of the small insects which they eat.

A great deal has been found out about the process of animal development by experiments on newt embryos. A small piece may be cut away and grafted onto another embryo or to an abnormal part of the same one. The earlier this is done the more likely it is to develop like the tissues around it. But if it is done later, the fate of the tissue is already determined, and one gets a newt with five legs or two tails. Similarly by tying a thread round the egg so as to dent it without cutting it in half one can get tadpoles with two heads and four arms, or two tails and four back legs. Such experiments have disproved two theories about animal development. One is the crude mechanistic theory that each part of the egg is destined to make a particular part of the tadpole, and that development is a mere unfolding. The other is the theory that the animal's soul, or some other such agency, imposes form on a crude and unformed material. In fact the different parts influence one another in an extremely complicated manner, but each may develop in different ways according to the influences acting on it.

However, there is an immense amount that we do not know about newts. As the males certainly do not seize the females by force, and do not seem to fight for them, their success or failure as fathers must depend on their skill and persistence in dancing, and on their bright colours. According to Darwin's theory of sexual selection these habits and colours have developed because those males which excited the largest number of females transmitted their characters to most progeny. There is evidence for the truth of this theory in some birds, and evidence against it in others.

But so far as I know no one has checked it up on newts, though they are easier to keep and watch than birds. There is room for one or two small aquaria in many working class homes. Accumulator jars do very well. The British species can be crossed artificially; but no one knows whether the hybrids are fertile, like the hybrids of the dog and wolf, or sterile, like those of the horse and donkey.

In the Soviet Union aquaria were very well developed. The finest exhibit of species with great differences between the sexes that I have ever seen was in a Moscow aquarium. In 1934 it was easier to buy small tropical fish in Moscow than in London. A biologically minded worker cannot have his own zoo. But he can and should have his own aquarium.

Cats

We know roughly how many adult dogs there are in England, because apart from sheep-dogs and a few others, they are taxed. We do not know how many cats there are. Recently Mr. Matheson, of the Natural Museum of Wales, got the help of a number of school children to count the cats in certain areas of Cardiff and Newport. Each child reported the number of adult cats in its home.

He concludes, from these and other statistics, that the number of cats in an area is roughly proportional to the number of human beings, and not even roughly to the number of acres. The number of cats with a fixed abode is about $10\frac{1}{2}$ per cent of the number of humans, the number of strays about $2\frac{1}{2}$ per cent. So there are probably about five million cats in Britain, apart from young kittens. However, the number of cats per hundred human beings is nearly three times higher in the slums round Cardiff docks than on the housing estates. A cat on a housing estate may be a luxury, but in some slums she is absolutely needed to keep down the mice and rats. So if we get the houses we want, the number of cats in England is likely to go down. For it is now possible to build a house so as to give mice nowhere to live, and to make cupboards mouse-proof.

I am interested in cats for a special reason. The colour and length of their hair varies a great deal. But they are all of much the same size and shape, apart from an occasional short-tailed manx. There are no peculiar shapes like the greyhound and dachshund, no giants like the carthorse, or dwarfs like the Shetland pony. Further, their matings are mostly governed by their own choice, not ours. So the cat population is much more like a human population, where a fair tall woman can marry a short dark man if she wants to, than it is like a population of dogs, sheep, or horses. Hence a study of inheritance in cats will be more help than a study in dogs to understanding inheritance in man.

At least one of the colour differences in cats seems to have been originally a racial difference. The wild cats of Scotland and

Europe are generally tabbies. But the cats shown in ancient Egyptian paintings, and those whose mummies have been found, were almost all yellow, or as they are commonly called, ginger cats. On the other hand I know of no evidence that there is or ever has been a race of cats all of which are black, or blue, though I am told that blue cats are much commoner both in Palestine and in Brittany than in England.

Things are much the same with man. There are countries like West Africa where all the native inhabitants have short hair and dark skins, others like England where all have long hair (if it is allowed to grow) and light skins. But other quite common characters are never characteristic of a whole race. Thus no one has ever found a race all of whom had red hair and freckled faces, though it would be easy enough for a führer who was a "man-fancier" to breed one.

We know a lot about how characters are inherited in cats, but not enough. There are two kinds of cat which I want badly. One is a tortoiseshell male (uncastrated, of course). Tortoiseshell females are common, but males are rare, and we do not know what character their children inherit, or why, as is often the case, they are sterile. The other kind is an albino, that is to say a white cat with pink, not blue or yellow, eyes. I can guess how it would breed, but I am not sure. If any readers can get me either kind, I shall be most grateful, and quite prepared to pay. But please write before sending any cats!

Why do we find it so much easier to make friends with cats and dogs than with other mammals of about the same size? The dog has a strong inborn tendency to social behaviour, learns to obey orders, and even develops something like a conscience. But the cat is not very social, and has little signs of conscience.

One reason is that cats and dogs have sensations much more like our own than those of hoofed animals such as horses, cattle, deer, and pigs, or rodents such as rabbits and rats. There are areas on the outside of the human brain concerned with sensations, not only from the eyes, ears, and so on, but from all parts of the skin.

We know this in several ways. Injury to a part of such an area does not abolish all sensation from the corresponding skin area, as when the nerves from it are cut. But it destroys its detail, so that the patient cannot say whether he is being touched at one point or several, or distinguish a penny from a matchbox by touch. And if the brain is exposed during an operation, then if the patient is conscious, stimulation of the appropriate part gives rise to sensations felt in the corresponding skin area; while stimulation of the skin causes electrical oscillations in the corresponding part of the brain, even in an anaesthetized man. In a cat, dog, or monkey, we can use this last method to determine the areas in the brain concerned with skin sensation. The animal is anaesthetized before the brain is uncovered, and killed under the anaesthetic; so it feels no pain. The results are similar to those in man.

But in many other animals Professor Adrian has found that most of the skin is not represented by sensitive areas on the rind of the brain. In a sheep, for example, only the mouth and feet are so represented. The pig has a large area for its sensitive snout; as a man does for his hands, which have as big a part of the brain at their service as the skin of the whole trunk. In consequence a sheep or pig gets very little detailed information from most of its body, while a cat or dog does. A cat likes being stroked. To please a pig you have to scratch it with a hard stick. A cat or dog can be gentle with its whole body. A horse can only be so with its sensitive muzzle. So dogs and cats can play with us, and we with them. In fact they play with children very much as equals, and quite understand that they must not use their full strength.

Some relatives of the domestic cat, such as the Scottish wild cat, certainly show no tendency to gentleness of behaviour, but others do so. The puma "Bill" at the London Zoo before the war, enjoyed tearing up newspapers, but would hold one's hand in his mouth without biting it. Unfortunately he was intelligent enough to understand that trousers do not feel, so he ruined a pair of mine without hurting the leg under them. I have little doubt that pumas could be made as safe domestic animals as our large races of dog. The most hopeful of all is probably the North American skunk. This animal defends itself when attacked by making a smell which will paralyse a man or a dog. It only bites in the last extremity. So if the scent glands are removed, which is not difficult, it makes an excellent pet. There are, in fact, probably quite a number of wild species which could become as good friends of man as our cats. If they are not domesticated before the spread of agriculture wipes them out, the loss will be irreparable.

Flying Ants

A few days ago winged ants were swarming in the outskirts of London. Now they are probably swarming in the Midlands, and will be so in Scotland in August. I did my best to stop a little boy from stamping on them. He said his mother had told him they were biting flies.

Actually they are the sexual forms of ants. The ordinary ants or ground staff, generally called workers, are females which have never developed sexual organs or wings. The winged ants are fully developed males and females. After mating, the females lose their wings, often biting them off, and try to found new nests. Naturally the vast majority fail. Those that succeed stop working as soon as they have a brood of worker children to look after them, and spend the rest of their lives laying eggs.

Thus an ants' nest, like a hive of bees or wasps, is normally a single very large family with its mother, sometimes a mixture of two or more related families. But there are larger differences between its members than between members of any human family. The most obvious difference between human beings of the same family is that of sex, and this is determined long before birth. Similarly the difference between fully male, female, and worker ants is determined while they are still grubs.

We are apt to take the sex distinction as something fundamental and all-pervading among animals and plants. It is not. Many animals, for example earthworms and most land and fresh-water snails, are hermaphrodite, combining both sexes. Some of them need a mate; others can mate with themselves. In many animals one generation is hermaphrodite, the next has two sexes. Sometimes, as in plant lice, one generation a year consists of males and females which must be mated. The others contain only females which do not need mates. Moreover, other differences can be just as important as sex.

In bees and wasps it is fairly sure that the workers fail to develop sexually simply because they are undernourished. This has been repeatedly shown by giving the special food intended for future queens to grubs which would otherwise have grown up to be workers. The sterility of worker bees is also probably due to malnutrition.

These insect communities have naturally attracted students of human society. When it appeared to be stable, as in the Middle Ages, they were held up as models, the mother being called the queen, or even, as in Shakespeare's *Henry V*, the emperor. To-day they are used as warnings of what will happen to men if they adopt socialism. Both these analogies are false. An ants' nest or a beehive is not a state, but a family. It has no government. There is no private property, although under socialism the average citizen will have more, not less, private property than to-day, besides his or her share in the public property.

A society of the insect type is impossible to man for many reasons. We do not lay eggs, nor bring forth large litters. Hence to keep the population up, most women must have children, and reproductive specialization is impossible. Malnutrition does not produce a special type of human being, but merely an unhealthy one.

Again in an insect society there is no specialization in social functions other than sex, except where there are anatomically different types, such as "soldiers" with large jaws. The same individual will do nursing, food-gathering, building, and fighting, at different periods. The only known exception is in bees, where each individual specializes on a particular species of flower.

Hence insect communities have never developed a class society on the one hand, or on the other a society where different members have their own skills, but each respects the work of his fellow, and all take part in directing the life of the community.

But the greatest difference is probably that there is no tradition in insect societies. Language is rudimentary. Bees can communicate. One type of "dance" means "I have found honey," another "I have found pollen," a particular smell means "Come here."

But these are no more a real language than the cries of birds. The "queen" does not educate her young even to the extent that birds do. On the contrary, an insect society works on a basis of inborn responses. The grubs give their nurses drops of a sweet secretion, and are fed in return. But the worker ants will also feed other insects which will give them sweet juices, even if these eat the ant grubs. So innate responses are liable to lead to antisocial conduct, as in men, and are far harder to modify by experience.

These innate tendencies are certainly not due to the inheritance of habit. If they were, the workers would have the "instincts" of sexual individuals, since all their ancestors were fully developed females or males. Lamarck thought that instinct was inherited memory. Thus newly hatched chicks were supposed to peck at corn because their ancestors have found seeds edible, and spiders to make their webs because their ancestors have gradually learned to do so. If this were true the instincts of worker insects would gradually come to resemble those of males and females.

Social insects can only change their habit by evolutionary change, which is a very slow process indeed. But human societies, even the most conservative, change very quickly when judged by the time scale of evolution.

Lamarckism is a socially harmful doctrine, though not so harmful as the Nazi race theory. For example, the late Professor Macbride, who also spread Nazi propaganda in Britain, wrote that Indians could not govern themselves because it took many generations of gradual self-government before a people could develop the necessary inborn qualities. Similar arguments are produced to justify "aristocratic" government. They certainly could not be used to justify either capitalist control of the State or the present House of Lords, since few rich men of to-day had rich grandparents, while many peerages are of recent date, and the older creations have taken so many brides from the stage and from newly rich families that they have not very much "noble blood."

Human behaviour depends much more on environment than ancestry. That is why it is possible to bring a people from capitalism, or even feudalism or barbarism, to socialism and democracy in one generation. The ants are stuck in their state of society, and we are not. But that is no reason for stamping on them.

Bird Migration

As we looked up to see the Fortresses going over to bomb airfields and factories in France, we saw another section of the cross-channel air traffic, and our song birds and swallows going to warmer countries for the winter.

The main routes of migration are roughly known. The long distance record is held by some of the swallows, which winter in south-west Africa. This is a distance of about 5,000 miles—actually more, as the birds do not fly straight. Of course they alight on the way.

It is not only the birds which migrate. This summer[1] I have seen a Painted Lady and several Clouded Yellow butterflies. As they never winter over in England, though they can breed here in summer, they or their parents must have flown over from the Mediterranean coasts of Europe, or even from Algeria, where they can live in winter. Only a few of the butterflies of any species in France cross the channel. But there are some species which migrate in masses like birds. The American Monarch or Milkweed butterfly regularly flies north from the southern United States as far as Canada, in the spring, and some at any rate fly south in the autumn. A few members of this species are caught in England; thirty-three is the biggest number in one year, but whether they fly the Atlantic or hitch-hike on ships is not quite certain.

Similarly a few birds migrate in a very irregular way. For example, every twenty-two years or so Pallas' Sandgrouse arrives in England from Siberia in small numbers, and we occasionally get birds from the Arctic or tropics.

We know something about why birds migrate, but nothing about how they find their way. The most important work on the causes has been done by Professor Rowan in Alberta. He showed

[1] 1943.

that even canaries, which are native to warmer climates than ours, can live out of doors in many degrees of frost if they are well fed. But they need a lot of food to keep warm. So he believes that our migrants have to leave because of food shortage rather than cold. What makes them leave is neither cold nor hunger, but the shortening of the day. He kept crows of a migratory Canadian species in a large cage which was floodlit every evening in the autumn, so that the effective length of night did not increase. He found that when released, most of them flew north instead of south, like birds which had had a normal series of nights of increasing length. A number of plants also react to the shortening of the day. Thus many trees only shed their leaves when the days draw in. Walnut trees usually die in Leningrad because the frost nips them before the leaves fall. They will live if covered with a tarpaulin about 3 p.m. early in September, in which case the leaves fall before the first hard frost.

The longer nights act indirectly by making the birds' ovaries and testicles diminish in size and cease to secrete hormones. They do not fly south if the appropriate hormone is injected. And the urge which makes them come back to their breeding places is due to the growth of the same organs in spring. Castrated birds do not migrate regularly. In fact the influence which makes our birds return in spring is the same which later on makes them desire to mate and build nests. We must be very careful in attributing human motives to animals. But the emotion behind migration to breeding places is almost certainly more like human love than hunger or curiosity. The robin is a good example of the exception which proves the rule. It does not usually leave us in winter; and in the autumn its ovaries and testicles increase in size, and produce enough hormones to make its breast redder in winter than in late summer, and, what is more, to keep it at home.

We do not know the answer to the most interesting question, namely how migrants find their way, and particularly how, in some species at any rate, young birds migrate in the right direction without any teaching. This kind of question is commonly called a mystery. I don't like this word. It is taken over from the vocabulary of religion, where it means either something not to be disclosed to the general public, or something which human reason cannot understand. This is just one of the uncounted number of problems awaiting scientific solution. These problems get solved in the long run; for example, we have solved the problems of where the swallows go in winter, and how bees communicate information, both of which baffled our ancestors.

Very likely when we discover the answer it will help airmen to find their way in darkness or fog. At one time it was thought that short wave radio disturbed birds in their flight. I am not disclosing a military secret by remarking that if this were true few birds would find their way to or from England during the war. Nor do magnets put them off their course. Meanwhile we want a lot more information. Before the war thousands of Soviet village schools were doing a co-operative study on bird migration. They caught migrating birds, put rings on their legs, and released them to be caught again elsewhere. No doubt these children, if they have not been killed or enslaved by the Nazis, are much too busy now. But they or others will start again.

This is one of the problems which is as likely to be solved by ordinary people in their spare time as by laboratory scientists. With the combination of scientific education and leisure to which we may look forward as Leninism spreads over the world, we can look forward to a day when about one person in twenty will be a naturalist, and many mysteries of nature will be mysteries no more.

The Starling on Trial

A Soviet scientific film of which I recently helped to translate the words, showed a remarkable experiment carried out by school children. A starling had nested in a special box. Its nestlings were removed, and a wooden model of a young bird substituted. When the mother perched on the edge of the nest the dummy opened its mouth and was fed. The food was collected in formalin, and at the end of the day the children examined it. They found that it consisted mainly of caterpillars, beetles, and other insects harmful to crops and trees. The moral was that starlings should be encouraged as friends of the farmer.

The film is well worth seeing, but it may not tell the whole truth. In a paper recently read to the Royal Society, Dr. W. S. Bullough, of Leeds University, accused the starlings of carrying foot-and-mouth· disease from Europe to Britain. This disease affects cattle, sheep, pigs, and a number of other animals, and very rarely, men. The animal gets fever, and blisters break out on the thinner parts of the skin, including the udders as well as the feet and mouth. Most animals recover, but they cease giving milk, lose weight, and may abort or go lame; so there is considerable loss. The disease is extremely infectious, and though in most countries the sick animals are allowed to recover, in Britain they are killed and burned. Most outbreaks can be traced to another in the neighbourhood. The agent of the disease, which is a virus too small to be seen with the ordinary microscope, can be carried not only by infected animals, but on men's boots.

At most times there are no cases in Britain, and then a farm will suddenly be infected, and an epidemic may start. Between 1900 and 1937 there were 349 unexplained outbreaks. It has frequently been suggested that birds are the carriers, and as starlings not merely feed in pastures, but perch on cows' backs, they come in for special suspicion.

For many years ornithologists have studied the migration of birds by fixing rings on their legs and then trapping the ringed birds in other countries. Thus we know that some British swallows go to South Africa in winter—our winter, that is to say. Starlings don't go so far. Some stay at home. Others leave for Europe in March, and come back about September to stay the winter. The Scottish migratory starlings mostly go to Norway for the summer and breed there. The English birds go to Sweden, Finland, the western parts of the Soviet Union, Poland, and Germany. The bird shown in the Soviet film may have wintered in England.

There is very little foot-and-mouth disease in Norway, and only ten unexplained outbreaks occurred in Scotland in thirty-eight years. The peak month for them in England is October, and they occur earlier in Eastern than Western England. On the other hand, in Sweden most outbreaks occur in April just after the starlings arrive, and they fall off in the autumn.

A further piece of evidence comes from the habits of the starling. Between June and December these birds live in huge communal roosts in woods and reed beds, or on buildings such as St. Martin's church in Trafalgar Square. In the spring they build nests and live in pairs. But except from March to May the roosts are fairly crowded. Less than 300 starling roosts are known in the whole of Britain, and over 50,000 birds may crowd together in one roost. The distribution of the roosts and of the unexplained outbreaks of disease on the map are very similar. Dr. Bullough thinks that the starlings, which sleep touching one another in the crowded roosts, may pass the disease germs to one another, and thus spread it. Except for the outbreaks when the continental birds come back the incidence of the disease agrees very well with the counts which have been made of the numbers of birds in roosts.

The evidence against the starling is only circumstantial so far, and some workers on the subject are very sceptical of Dr. Bullough's theory. The answer is likely to come from a much more detailed study of the movements of starlings, which may fly as far as fifty miles a day to and from their roosts. In the Soviet Union research on bird migration was done on a big scale by school children, who caught birds, ringed them, and released them to be caught in turn by other school children. In England such research is left to adults with spare time. I think that at least one child in ten has a genuine interest in animals, and could become a naturalist. But a bright boy with an interest in biology is encouraged to learn up the rabbit's anatomy with a view to a scholarship, rather than "waste his time" watching wild rabbits.

Even in towns a good deal of natural history can be done. To take one example, as many as fifty red underwing moths have been seen on a single London lamp post, and such animals as the mouse, clothes moth, and house fly are, of course, commoner in towns than in the country. If we are to make the most of Britain as a source of food, wool, and timber, we must study its wild animals and plants, as well as the domesticated ones. Nature study will come alive when our country belongs to all of us, and every citizen feels that it is up to him or her to make it more productive and more beautiful.

Instinct

A correspondent has sent me the following question: "What is instinct, and how far does it differ from intelligent reasoning? To what extent is it possessed by (*a*) the lower animals, (*b*) the human race?"

The word instinct was invented to "explain" the behaviour of animals at a time when they were believed to be utterly different from men, in having no reason, and no souls. It is very little used by biologists today. It denotes an inborn tendency to do certain actions, often quite complicated, in suitable circumstances. Now we do a great many very complicated things without thinking about them. For example, after swallowing our food we bathe it in various digestive juices, churn it in our stomachs and guts, passing it down as each stage of digestion is complete, and then absorb some parts of it and reject others. Animals do the same. But we do not call this instinctive behaviour. It is a series of reflex actions, whose mechanism was studied by Pavlov among others. We reserve the word instinct for actions of a kind which in ourselves are conscious and willed, and may be reasoned.

Numerous animal actions often described as instinctive are mere reflexes. Many insects fly or crawl to a light, and have been said to desire it. If we observe one of them we note that if we shine the light on its right eye, it pushes more vigorously with its left legs or wings, and turns right until it is facing the light. So would we if we wished to walk towards the light. But now take one of the supposedly light-loving insects, and put black paint on its left eye. It will continually move as if the light were shining on its right eye only, that is to say it will go round and round in a circle, always turning right. Activities like these are properly called reflex rather than instinctive. Our own efforts to keep our balance are of this kind. We can easily throw them out of gear by walking round rapidly with our foreheads on a walking stick, and then suddenly standing up.

We use the word instinctive for behaviour which is not so mechanical as this, without its objects being fully understood. A female cat may have been taken from her mother at birth, and fed from a fountain pen filler, so that she can have no memory of motherhood. But when she bears kittens she licks them, suckles them, keeps them warm, carries them to a nest, and so on. She probably loves the kittens and certainly enjoys touching them, but it is idle to pretend that she knows that she can give them food, and that otherwise they will die. You might as well suggest that a child wants to eat sweets because it knows that it needs chemical energy to keep it warm.

A human mother cannot do as well as a mother cat without teaching. Her instincts are not so precise. But if properly taught, and provided with the necessary food, clothes, housing, and so on, she can do much better, as is shown by the fact that in progressive countries the mortality among babies is less than among kittens.

When some precise pattern of behaviour is universal in a species, we are apt to call it instinctive. This is sometimes, but not always, true. We can only decide by experiment, and the greatest number of experiments have been done on the song of birds. This may be instinctive. For example, a male blackbird, whether brought up alone, or where he can hear other birds, only sings the blackbird song, which is fairly complicated. But the chaffinch has to learn his song. If he is not taught it, his song is described as like that of the lesser whitethroat. Other birds will learn the song of a different species, even from a gramophone record.

It is clear that men and women are more like chaffinches than blackbirds, but not very like either. We learn the language of the family or orphanage in which we are brought up, but children brought up by animals do not develop a language of their own. Babies have a certain urge to speak, but it is not well enough developed to be called an instinct.

We are very apt to think that sentiments held by the vast majority of men and women are instincts. For example, most people object to murder within the community, even if they commend the murder of members of another tribe. But primitive societies have been discovered in which a murderer is not punished, but is expected to adopt the children of his victim.

Man has become the most successful of the animals because he is the most plastic as regards behaviour. He can learn to do a vast variety of things and, which is more remarkable, to desire a vast variety of things. An animal species adapting itself to a new environment must change its instincts. This is a slow process, like the change of form, occupying many thousand generations.

Human character can be changed in one generation. The younger generation in the Soviet Union mostly take it for granted that men and women will work together for the public good. They regard the struggle for one's own interests which is inevitable under capitalism, as being not so much wicked as ridiculous selfishness. The young Nazis and Japanese believe that they are members of a master race, and combine cruelty to foreigners with blind obedience to superiors. The battles which are going on today will decide which of these ideologies will be a model for all mankind.

We cannot draw sharp lines between reflexes, urges accompanied by desire, and instincts. But we can say that instincts producing highly complicated and stereotyped behaviour are most highly developed in insects, and that in man they are less stereotyped than in related animals. We have to learn most of our behaviour. And therefore we have greater possibilities for good or evil than any other animal. The domesticable animals, and notably the dog, have a large capacity for learning. Man differs from them, not only in having less instinctively fixed behaviour, but in modifying his environment by production, so that he is always having to learn new activities. If we had complicated hereditary instincts we should still be stuck in the old stone age, if we had even got so far. If we believe that all the customs of our own society are "human nature" we shall be unable to adapt ourselves to the great changes which are now upon us.

The Origin of Species

Men started naming different kinds of animals and plants long before history began. This is shown by the fact that some animals have similar names in languages such as Latin and the ancient

Indian Sanskrit, whose common ancestral language must have been spoken many thousand years ago. Primitive peoples whom we call savages—though that is probably nothing to what they call us—often have names for hundreds of kinds of wild animals. These names are obviously useful. Clearly the differences between two cats are less than those between any cat and a tiger. In the middle ages the philosophers whose teaching was accepted by the catholic church thought that the names stood for forms common to all members of a "species," and having a real existence of their own.

Linnaeus, the Swede who founded the modern system of classifying animals and plants, thought each species had been created separately. Lamarck, largely from a study of fossil animals, thought that they had been formed from other species in the past, but his theory of how this had happened was incorrect. Darwin produced much stronger evidence for the origin of species, and his theory of how they originated is much nearer the truth.

But it was not the whole truth. He pointed out that by selection men had produced races of dogs, pigeons, and other animals and plants which would certainly be put in different species if they were found wild. But his critics answered that they can still breed together. Even a Newfoundland dog and a dachshund have given fertile hybrids, whereas a dog and a fox do not produce hybrids at all, even if artificially mated, and hybrids between a horse and a donkey are sterile. It is true that some animals and plants of obviously different species will give moderately fertile hybrids, for example the large and small elephant hawk moths, and the European and Chilean strawberries whose crossing gave our cultivated varieties. But other species which resemble one another closely will not do so.

The work of the last thirty years has completely removed this objection to Darwin's theory, though it has shown that it has to be modified in another respect. Clearly the conclusive experiment is to start with a group of plants or animals which belong to the same species and breed together, and from this to make another group which can breed with itself, and not with the remaining descendants of the original group.

This was first done for a plant in London by Crane and Jorgensen working with the tomato, and for an animal by Koshennikov in Moscow with a small fly called *Drosophila*. If you repeatedly cut a tomato shoot back, some of the new shoots will have thicker leaves and other differences. If these are cut off and planted they will set seed with their own pollen, giving more but smaller fruits than the original. But they give very few hybrids with the original stock, and these are highly sterile, giving one or two seeds per plant at most. The change is due to a doubling of the number of chromosomes in the nucleus of each cell. Microscopical examination shows that a number of species have arisen in this way. These new species are generally rather less fertile than the parent, but stand frost better, so they are common in the arctic and in mountains.

Another way in which new plant species arise is by a doubling of the chromosomes in what started as a sterile hybrid. Thus species can arise at one single leap. Perhaps Darwin's political and philosophical outlook, which was that of the 19th century British upper middle class, gave him a bias in favour of slow change.

Still the differences which prevent crossing in most animal species have almost certainly arisen slowly, and we find all sorts of intermediates. For example, when many species are crossed, the hybrids of one sex only are fertile. Indeed the rule which generally enables one to predict which sex will be sterile, if only one is so, is called Haldane's rule, as I discovered it.

Some wild species seem to be in process of splitting up. It is not enough to form new varieties. If these mate freely, as the different colour varieties of our mice, adders, newts, snails and grasshoppers do, the species will merely remain variable.

But if different varieties have different habitats or breeding seasons, or show a repugnance to crossing, a species may break up. For example, the peppered moth and the mottled beauty have developed black races in the "black countries" of industrial areas, fairly sharply separated from the normal races outside. They still interbreed with them freely, but after a few thousand generations small alterations in the chromosomes of one or another would probably lead to partial sterility in the hybrids.

However, we are likely to follow the Soviet example and gasify most of our coal underground within a generation or two. If so, these black moth races will disappear again, with the black surroundings which they fit, before they have had time to form species.

Thus we see that Darwin was largely correct in his views as to how new species arose, but that like many other thinkers of the 19th century, he overestimated the "inevitability of gradualness."

Species in the Making

How many wild species of mammals are there in Britain? One would think it would be easy to answer this question, except that there might be a doubt about some species which have recently been introduced. The American grey squirrels, some of which were released from the London Zoo early in this century, have spread over most of southern England. A small Japanese deer is wild in several southern counties. The fat dormouse from France, and the American chipmunk, have small colonies in this country, but it is not sure that they will establish themselves.

Apart from these there are certainly thirty-seven wild species, and perhaps fifty-two. A hundred years ago most naturalists believed, with Linnaeus, that species had been created once for all, and that the number could be definitely determined. But Darwin had already begun to doubt this. Now most biologists think that species have arisen in the past, and are arising today. The difficulty as to the number of English species is due to the fact that some of our species are in the process of splitting up, and there is doubt as to how far the process has gone.

A number of our species are variable. The variation may be due to changes in environment. The fur of the Scottish Highland stoats changes from brown to white every autumn, and they are then called ermines. The stoats of southern England do not change, even in a hard winter. Or it may be determined by heredity. Black, yellow, piebald, white, and hairless freaks have been found in various species, such as mice, rats, squirrels, rabbits, and moles. These abnormal characters are generally inherited if the animals are caught alive and bred in captivity.

But such freaks are not regarded as new species or subspecies, nor even in most cases as incipient new species. They breed with the other members of their species, and nowhere do they form a majority. Probably natural selection prevents their spreading. This could only be proved by counting thousands of animals. This can be done on insects. For example, in the case of a small British fly which my department is studying, yellow forms occur, but never spread. This is partly because the females prefer the normal type, and partly because the yellows dry up more easily in a drought.

On the other hand there is sometimes good reason for splitting what was formerly regarded as a single species. For example, Linnaeus, the great Swedish classifier of the eighteenth century, distinguished the wood mouse, or long-tailed field mouse, *Apodemus sylvaticus*, from the house mouse. In case anyone thinks I am writing of high-brow and remote problems I may add that this is almost certainly the commonest British mammal, and that it can be a curse to allotment holders by digging up newly-planted peas and other seeds.

In the late nineteenth century the naturalist de Winton found that in southern and eastern England there is also a very similar mouse, with a yellow neck, which is slightly larger than the long-tailed field mouse, and has three more joints in its tail. He regarded it as a new species, and called it *Apodemus flavicollis*. Most naturalists probably agree with him. But since then three other species have been distinguished, namely, forms from St. Kilda, the Hebrides, and Fair Island, and also a subspecies from Bute. They differ from the ordinary wood mouse in colour and average size, but not as sharply as the yellow-necked, and probably have less claim to be a distinct species.

On the other hand they may very well be species in the making. The question of their specific rank can only be cleared up by breeding experiments. Here are some questions to be answered. Do the wood mouse and yellow-necked mouse breed together in nature? If not, can they be got to do so in captivity? If so, do they produce hybrids? Are these hybrids sterile, like mules, or partly

fertile, or fully fertile? If they are fertile, are the differences inherited as a single unit, like the difference between ordinary and
black rabbits, or in some other way?

If I had to bet on the answer, I should guess that the yellow-
necked mouse would not breed with the wood-mouse, or that if
it did the hybrids would be sterile, while the island forms would
breed with the ordinary wood mouse, and there would be some
blending in the hybrids, as is usual in crosses between species. But
I may well be wrong. On some of the Pacific islands, birds whose
ancestors migrated there from the continents or larger islands have
been isolated long enough to form distinct species which do not
interbreed with the mainland forms. Almost all our larger animals,
and probably our bats, of which there are eleven kinds, are divided
sharply into species which rarely if ever interbreed. For example,
the rabbit and hare, the weasel and stoat, the red and fallow deer,
are quite distinct. But the wood mouse, the bank vole, the field
vole, and possibly the house mouse, seem to be in the process of
forming new species, while Irish races of the stoat and alpine hare
are sometimes thought to be distinct species.

The theory of evolution was founded by such naturalists as
Darwin, Wallace, Bates, and Muller, who studied animals all over
the world, and particularly in the tropics. But they were observers
rather than experimenters. They found species which had obviously originated recently, for example wingless insects on islands;
and they produced theories, which seem to be substantially true,
about the origin of species.

But these theories can only be verified by watching species in
the act of splitting. All our three species which are engaged in
splitting are liable to such increases in numbers that they become
pests to agriculture. So active research on them will be of practical as well as theoretical value. Here is a job for British naturalists after the war.

Back to the Water

Much of my work during this war, some of which I have been
allowed to describe, has been in connexion with human life under
water. Naturally I had to see if I could get any hints from other

air-breathing animals which have taken to life in water, and about whose physiology something is known.

Both the study of fossils and that of comparative anatomy make it fairly clear that there were animals in water before they came on land. And they leave no doubt that the four-footed land animals are descended from fish which came out of the water about the time when the old red sandstone was laid down, before the coal was formed. Probably we land vertebrates are all descended from a single species of adventurous fish. But very many groups of land vertebrates have gone back to the water. Some, like water-voles, otters, gulls, and sea snakes, show no very obvious changes in their anatomy. But these spend a good deal of their life ashore, and above all their young are born or hatched in the air.

Other aquatic animals such as seals, penguins, and turtles, are so far transformed that their limbs are greatly modified for swimming, and neither seals nor turtles can walk far, while penguins cannot fly. But they still come ashore to bear their young or lay their eggs.

Only two groups of mammals have been fully modified for aquatic life, and live their whole lives in the water. These are the flesh-eating whales and dolphins, and the vegetarian sirenians, such as the manatee. No bird has ever managed this, and the ichthyosaurs, an extinct group of reptiles which did so, brought forth their young alive, which is fairly unusual in reptiles, but obviously necessary in an air-breathing marine animal.

The same kind of thing has happened in the evolution of insects. Their remote ancestors were aquatic, and many different groups have gone back to the water. Again this is usually only for part of their life cycle. Thus dragonflies, mayflies, caddis flies, and mosquitoes spend their larval stage under water, and do all their growth there. They only come out at their last moult, mate in the air, and lay their eggs on or near water.

In the course of evolution there are comparatively few examples of water animals adapting themselves to land, and a great many of land animals adapting themselves to water. And it is a striking fact that many of the land animals which have gone back are more efficient in the water than its original inhabitants. The largest whales are larger than any fish of the present or past. And they almost certainly swim quicker, grow faster, and are more intelligent. As they keep their temperatures steady, they can live in warm or cold water, and thus have a wider range north and south than any fish.

Clearly life on land has given them some useful characters which they have taken back to the water. A fish coming out of water finds itself in a more difficult environment than before. It is not supported on every side. It must develop limbs, which need much more complicated brain function to work them than fins. It needs stronger bones, and must give its young a tough egg shell or bring them forth alive. It is subject to a greater range of temperatures, and it is an advantage to it to be able to keep warm in cold weather and cool in hot weather. A constant temperature is probably needed for the high development of brain function found in mammals and birds. A "cold-blooded" animal, more accurately an animal of variable temperature, probably could not develop great mental powers. Certainly the human brain is more upset by a rise or fall of temperature than our other organs.

So the adaptations developed by land animals to meet the difficulties of life on land, of which I have only mentioned a very few, have proved useful to their descendants which went back to water, even though they continue to be air breathers, and must come up to the surface from time to time. Seals, porpoises, and probably whales, manage to dive for many minutes without suffocating, by cutting off the blood supply from most organs except their brains, and slowing down their hearts so as to conserve oxygen. Human divers cannot do this, so if they are to stay down even for five minutes they must have air pumped down to them, or take compressed air with them in cylinders.

The facts of animal evolution have a considerable bearing on the development of human societies, as Marx saw when he wished to dedicate "Capital" to Darwin. Of course, however, one must be careful not to argue uncritically from one to another. But the history of whales and the like may help us to understand why scientific communism could not develop directly from primitive communism, but a period of class society was inevitable. This is important if we are to see through the arguments of some anarchists, simple-lifers, and others who want us to discard many of the good things of capitalist civilization along with its evils.

A member of a primitive communist society is bound by custom, and seldom seems to think for himself, as Engels pointed out. Worse still, he may treat the members of his own tribe like brothers and sisters, but find his greatest pride in collecting the heads of a tribe five miles away. Apparently class society was necessary to develop the division of labour, to allow the formation of communities so large that each member does not know every other, and above all to develop technology beyond that of the stone age. In class society each man must fend for himself, and thus develop intellect, if not morality.

We now know that it is possible to keep these gains while abolishing the class distinctions which helped to generate them. There is no more reason to suppose that men in a scientific communist society will go back to the primitive mental and moral processes of primitive men, than that whales will become cold-blooded or give up suckling their young.

No Caterpillars by Request

In the *Worker's Notebook* for August 1st Walter Holmes said that the lobster is "a very primitive form of life." As a biologist and chairman of the editorial board I can't let that sort of thing go by, especially as it may be used to justify the Ministry of Food's permission to serve lobster, though not fish, along with meat in restaurants.

You've only got to look at a lobster to see that it is a fairly complicated sort of animal, and if you can afford to buy a whole lobster to cut open, you will see that some of the gadgets inside it are not so simple either. An ant or a fly is just as complicated inside, but you can't see their internal organs without a microscope.

By a primitive animal I suppose we mean an animal like those which existed a long time ago. For example, there have been limpets very like those which live today ever since the Silurian rocks were laid down in the sea which is now Wales, over three hundred million years ago. In fact, compared with a limpet, Mr. Chamberlain is quite progressive. However the limpet, though primitive, is not very primitive. It has a mouth, eyes on stalks, and various other organs. It has a lot of evolution behind it; but after evolving so far, it has stayed put for a long time. A worm, a jellyfish, or a sea anemone is much more primitive than a limpet.

We can tell a good deal about which animals are primitive by studying fossils. Fossils show us that fish were there before amphibians (like the newt and frog) which developed from them; amphibians before reptiles such as lizards, snakes, tortoises, and crocodiles; and reptiles before birds and mammals. So we can say without hesitation that, for example, reptiles are more primitive than birds, which is fairly obvious for other reasons. But a great deal of evolution took place before any recognizable fossils are found, so we can't always answer on these lines.

However, we can compare the lobster with other fairly similar animals, and ask if it is primitive. It is obviously a segmental animal, that is to say, built up of a number of sections one behind the other, each with a pair of appendages. Centipedes and milli-pedes have a great many segments, each with a pair of legs, and one segment is very like another, except at the two ends. But in the lobster a lot of the segments are fused together, and some of the appendages are walking legs, others mouth parts, stalked eyes, and feelers, whilst the biggest pair carries the claws. So it looks as if the lobster had developed from an animal more like a centi-pede, by specializing the appendages in different ways, and fusing the segments in the front part of the body. The few avail-able fossils support this view.

We are very apt to think of evolution as a process which has culminated in ourselves, as if all the animals could be arranged in order like boys in a class at school, with the human species as top boy. This is quite wrong. Primitive animals such as jellyfish and sea urchins have no head at all. Some worms have the begin-ning of a head, and the lobster's head is not definitely separate from its body. But three groups of animals have independently developed a head with eyes and a brain near the mouth. Generally there is an organ of smell on the head also, for example, our nose, and the antennae or feelers of insects. But hearing organs may be anywhere. For example, many insects hear with their legs.

The three groups with heads are the higher molluscs, such as snails and cuttlefish, the arthropods, including insects, spiders, and lobsters, and the vertebrates, including ourselves. The molluscs have not got very far, but the insects have done very well.

If an intelligent visitor from another star had come to our earth a million years ago he would probably have said that the insects were the most advanced animals. He would have found insect societies which were not mere herds, but engaged in co-operative production, for example, of honeycombs. If he had been a biologist he would have seen that insects could not grow much larger, because their system of breathing could not work in an animal more than a few inches long. And so they cannot develop brains large enough for very intelligent behaviour.

The social insects have been called communists because though they collect food and manufacture such products as the paper nests of wasps and the honeycombs of bees, these are not the property of any individual. But further research has shown that they are not communistic in the human sense. For example, the wasps within a nest have a primitive kind of private trading. A worker catches a fly, chews it up, and gives it to one of the grubs in the comb. The grub repays it with a drop of sugary juice. Ant larvae do the same. And various beetles and other insects living in ants' nests exploit the workers by secreting small drops of sweet liquid. In return they are given much more than its equivalent of other food, and are sometimes even allowed to eat the young ants. In just the same way there are generally parasites in primitive human communistic societies, for example wizards, "rain-makers," and witch-finders.

Further, these insect societies are not run on the principle "to each according to his needs." A beehive in spring contains one fertile female, the queen, and a great many sterile female workers, who are all her daughters. In spring most of her eggs are put in cells with ordinary food, and develop into workers. But some are given a special secretion called "royal jelly" containing a hormone which causes the grubs to develop into queens. In fact the workers are produced as a result of under-feeding. So a certain amount of what, in a human society, would be injustice, is inherent in these insect communities. And the workers cannot revolt and take control, because they cannot reproduce themselves. So the insects, although very highly developed in their way, seem to be at a dead end so far as social evolution is concerned.

May I conclude with a plea to readers? A comrade in Devonshire has sent me a small caterpillar, by now very dried up, which he hopes I will identify. I could probably do so at the Natural History Museum in the course of an hour or so, but I perhaps flatter myself in thinking that I have more important things to do. I doubt if I could have identified it without help even before it changed colour on drying up. So may I respectfully ask my readers to remember that I am not omniscient, and am very busy? I wish I had time to identify caterpillars, which, by the way, are far from primitive. In fact a caterpillar has more different muscles than a man. And the Chinese regard it as a great delicacy. But in spite of this, please send me no more caterpillars.

Domestic Animals

What was the most important event in man's past? I do not say in human history, because, if I am right, it took place before any historical records were made, and even before the origin of the rather dubious legends with which history begins. At first sight one would be inclined to choose some great political transformation, such as the fall of the Roman Empire or the Russian Revolution, or else the beginning of some world-shaking system of ideas, such as the origin of Christianity or the writing of the Communist Manifesto.

However, if the Marxist account of history is correct, these events, or something like them, were bound to have happened. The Roman Empire was based on slavery, and was bound to break down or change very greatly when slavery ceased to work. The New Testament is the only set of books surviving from the

Graeco-Roman civilization which were written by workers (except perhaps for St. John's gospel), for workers, and about workers. That is one reason why they have had an influence that no other books have had. There were probably other books written by workers at the same time, preaching more revolutionary doctrines. But they have not survived. Some organized ideology for the Roman workers was a historical necessity. The Roman government even went so far as to invent a special religion for freed slaves, but it did not get very far! In the same way, once the industrial revolution had produced a proletariat, socialist thinkers were bound to arise, and many did. But it was Marx and Engels who laid the theoretical foundations of the only Socialist State.

We must look behind changes in government and ideas to changes in productive forces and relations, and of these, the changes in productive forces are the most fundamental. The first and greatest of these was the origin of production. The earliest fossils that are at all human are associated with tools, if only with crudely chipped stones; and Engels was probably right in saying that an ape-like creature became man when it started making tools. If so this change was the origin of man, not part of his history. Men continued for hundreds of thousands of years as hunters, using fire, stone axes, and later bows and stone-headed arrows. They developed sculpture, painting, and dancing, made huts, and probably wore skin clothes in winter.

Towards the end of the old stone age, dogs joined human society, probably as scavengers. From what we know of the habits of related species, such as jackals, and of the relations of dogs to primitive human societies, it is likely that the initiative came from the dogs, and it is fairly sure that at first they were not private property.

But, at least in Europe, there was a rather sudden change about 10,000 years ago. Men began to domesticate animals and probably to practise a primitive agriculture, and changed over from the use of chipped to polished stone. The domestication of animals or plants as food sources brought about vast economic changes. To take only one of them, the number of people who could live on a square mile of fertile land was increased from ten to a hundred times. This meant that society could no longer be based on a small tribe of twenty to a hundred people who knew one another intimately, and settled most disputes by common sense.

Some sort of government was needed, though not necessarily anything that could be called a state. But domestication meant the possibility of private property not merely in tools, clothes, and huts, but in sources of food, in fact the private ownership of land and capital. Many primitive tribes avoid the development of a class system based on wealth in ways which seem strange to us. For example, in some Pacific islands the produce of the garden worked by a man and his wife goes to his sisters and their children.

Once a man owned sheep or other sources of food, he could and did hire workers without property. The book of Genesis tells the story of how Laban hired Jacob for seven years, and swindled him at the end, while Jacob got the better of Laban at the end of another seven, and became a large-scale sheep-owner. In this kind of way class divisions in society arose. "The lamb misused breeds public strife," though not exactly in the way that Blake meant. Class society is quite a recent development, and in no way necessary for human existence or progress.

In fact the men or women who made the great step of domestication lived in a classless society. They must have been people of great intelligence. The idea of keeping a herd of animals with a tribe, instead of hunting it, was a very original one. It may have started in a people with a passion for pets, who began to eat them when they grew too many. It is certainly striking that even in early historical times the Greeks always sacrificed an ox or sheep to a god before eating it themselves, as if they were ashamed of the action, and needed a religious excuse. Even today good Muslims will not eat an animal unless it has been killed in the name of Allah, and many Hindus will not eat animals at all.

Galton believed that only a few animals had instincts which fitted them for domestication, and that all the large domesticable animals had already been tamed. There is much truth in this, as can be seen from the case of the budgerigar, which was only brought over from Australia last century, but is obviously better adapted for human society than any of our British wild birds.

However, animal behaviour can be altered by many generations of domestication; and the differences are inherited, as has been shown by crossing wild and tame mice. Today a number of fur-bearing animals such as the silver fox, the mink, and in the Soviet Union the sable, are being domesticated. And their descendants will probably be tame in a few centuries if people continue to wear furs, which is by no means certain, if only because synthetic fibres will probably replace them, as they are replacing silk.

It is very striking that soon after the domestication of animals, art almost disappeared. Neolithic carving is very crude compared with that of the old stone age, and painting is unknown. This may have been due to the origin of classes in society, and a consequent contempt for manual work. A scientific study of what happens when an animal is domesticated will not merely be of interest to biologists, but will help us to understand one of the greatest changes in the past of our own species.

What to do with the Zoo

For the first time the London Zoo has got a scientific director, Dr. Hindle. In former times the Secretary attempted to combine administrative with scientific work, but there is much to be said for employing a full-time scientist. Dr. Hindle has a very wide experience of foreign animals, for he has studied a variety of tropical and sub-tropical diseases, including yellow fever and kala azar, a malady which has killed millions in India, and which spreads at least as far west as Syria.

Both are insect-borne, and the best way of controlling them is to keep down the insects which carry them. But both of them are also diseases of other animals besides man. Some monkeys get yellow fever; and hamsters, which are little burrowing creatures half-way between a rat and a guinea-pig in appearance, get kala azar. These animals act as a reservoir of infection from which human beings can be infected, even after all the human cases in an area have been cured or have died. So a good deal must be learned about these animals before the disease can be controlled, and Dr. Hindle had to study them as well as insects.

Dr. Hindle will not be able to do very much for the Zoo at present, because of the impossibility of importing animals, and the shortage of staff, food and fuel. But there is one thing he could organize during the coming year, and that is an exhibition of all our British land vertebrates except those birds which migrate or need a lot of space to fly in.

For example, we have three native species of snake. The adder or viper is the only poisonous one. Vipers do not do very well in captivity, but will live several years if they are not roughly handled while being caught. The grass snake, on the other hand, which is also common, does quite well, and everyone should know the difference between the two. The smooth snake is a good deal rarer, and almost, if not quite, restricted to Hampshire and Dorset. The slow-worm or blindworm is not a snake, but a lizard which has lost its legs; as appears among other things, from its having eyelids, and its habit of breaking off its tail when alarmed. All these ought to be on show, if only to discourage hikers from killing the harmless ones.

The hardest English land vertebrates to get would be the pine marten, which still lives in the Lake District but is very rare, and the polecat from the Welsh mountains. But I think Dr. Hindle might be excused from keeping a mole. The mole is only happy underground, needs about fifty worms a day, and is said to die of starvation in a few hours if he runs short of them.

Even during the war we might well have a house for British animals only. It would not require heating, as they can all stand our winters; and the deaths of foreign animals have left the necessary room. After the war there will be great opportunities to make the Zoo a place of scientific teaching and research, without lowering its entertainment value. Here a lot will depend on co-operation with the staff. All of them are interested in their jobs, and some are capable of real scientific research.

Last year three of the keepers made history by publishing, in the Zoological Society's *Proceedings*, a detailed account of birth and infancy of chimpanzees. One mother brought up her baby without help; another abandoned it when it was born, and the keepers had to look after it, though later she took some care of it. Many of the keepers notice previously unknown facts about the animals in their charge, and during the war most of them must have made discoveries about what food animals will eat, for example, turnips instead of bananas. These discoveries should be made part of the general stock of knowledge.

In the past, the Zoo's most important contributions to research have been on the anatomy of the animals which died, and on their parasites. The anatomy could certainly be improved if it were studied from a more physiological angle. For example, a giraffe can stretch up to 17 feet, and its heart is about 8 feet above its toes. This means that the pressure of the blood in its leg veins and arteries must be far greater than in a man or horse. I am willing to bet that a giraffe's leg veins will be found to have a special structure if anyone examines them carefully next time one dies.

But even more important is the teaching side. It would be possible to state the class, order, family, and diet of every animal on its cage or paddock, and these facts would gradually soak into the public. We should come to take it for granted that the porcupine is a rodent like the rat and squirrel, while the hedgehog is more nearly related to the mole and the shrew. Only after we understand such classification can we really begin to examine the evidence as to man's place among the animals, and the arguments for and against evolution. If anyone thinks this unimportant, please remember that Stalin was sacked from a theological seminary for reading Darwin.

Still more valuable would be a link up with the naturalists of London, and especially with the boys and girls who collect caterpillars and newts, and want information about what they have found and how to keep it. Dr. Hindle could do worse, if he has not done so already, than spend an afternoon at the Horniman Museum at Forest Hill, which does a great deal for the young naturalists of South London.

London has a very big bird and insect fauna. I have seen a kingfisher within 50 yards of the main line from Euston. In the last two years two friends have caught a small elephant hawk moth and a chalkhill blue butterfly in their London gardens, though the latter is not supposed to live nearer than Berkshire. Other insects, for example, the gray dagger, sycamore, poplar gray, buff tip, and vapourer moths, are commoner in London than in the country. Few of us can keep a kinkajou or a badger. Most of us can keep moths or minnows, at least in peace-time, if we want to. By helping those who wish to devote some of their leisure after the war to the study of our native animals, Dr. Hindle can do a great deal to democratize British biology.

Spring

The primroses, almonds, and a few other flowers, are blooming in Southern England. The mild winter has of course helped them on, but why do they bloom when they do? That is the sort of question which scientists ask, whereas the plain man takes the course of nature for granted. But as most children ask questions of this kind, I don't so much think that scientists are an abnormal set of people, as that they are people whose normal curiosity hasn't been knocked out of them.

Of course it is not enough to ask such a question, or to think out an answer to it. It can only be answered by experiments, that is to say by putting the question to nature in actions, not words. You have got to alter natural conditions, and see what happens. One can try to imitate the various features of spring one at a time, for example by warming up plants in December, or giving them more light. One can't learn much about nature except by trying to change it. The same is true of human society.

That is why numbers of people who have tried to change some features in our existing society, whether it be the standard of housing, the criminal law, or the prejudice against coloured people, discover that the nature of society is what Marx and Engels said it was, and find themselves first working with the Communist Party and then joining it.

A great deal of our knowledge about what happens to plants in spring comes from practical horticulture, and more recently agriculture. If flowers are "forced," to use the horticultural term, to bloom in winter, they sell for a great deal more than their price in season.

The Soviet botanist Lysenko distinguishes two stages in the

development of annual plants, the first in which it is mainly controlled by temperature, and the second in which its making is controlled by light. Many perennial plants pass through similar stages. Although most seeds require warmth to germinate, many will not do so without preliminary exposure to cold. Thus apple and pear seeds should be cooled down to about 40° F, and gooseberry seeds to 20° or even 10°, that is to say twenty degrees of frost. Many varieties of wheat do not need cold to germinate, but unless they have been exposed to cold, produce only leaves, but no flowers or grain. Other cereals, such as millet, can be grown in cold climates if the seeds are given a week or so at about 70° F. This process of forcing seeds by heat or cold is called Yarovizatsia or Vernalization.

Cold and heat have similar effects on perennial plants. Nettles start to grow very slowly in spring unless their roots have been exposed to frost. On the other hand, if one branch of a lilac is placed for twelve hours or so in water at 85° to 90° F., it will produce leaves and blossoms before the buds have opened on the other branches.

Besides these methods, which copy nature, chemical methods are used. For example, lily of the valley or lilac can be forced to bloom by putting it in air-tight boxes exposed to ether vapour. In the United States seed potatoes are harvested in the north in autumn, and sent south in railway vans sprinkled with ethylene chlorhydrin. They are planted in the coastal states round the Gulf of Mexico, and start to put out shoots without any rest period, whereas untreated potatoes lie dormant for months.

In the light-sensitive period the behaviour of plants depends on the length of the days and nights, as shown by Garner and Allard in America. Many plants of temperate climates, such as wheat, barley, and oats, develop quicker the longer the day. By using powerful electric lamps at night one can get two or even three generations in a year, provided the seeds are cooled down. But many temperate plants will not flower at all unless the daylight lasts for over twelve hours. So in the tropics, where all days are about twelve hours long, they merely produce leaves, but no seed, even in cool mountainous regions.

On the other hand, plants of tropical origin often need a long night. They include cereals such as rice and millet, and many varieties of cotton and beans. They will only flower in climates with long summer days if they are covered up for some hours in the morning or evening.

Different varieties of the same plant differ both in their temperature and light requirements for early flowering. Lysenko and his pupils have produced extremely rapidly developing wheat by crossing two varieties, one of which responds well to cold, the other to long days. Even though neither set seed particularly early, some of their offspring combined the qualities of both parents. Thus Lysenko's work is being applied, not only to speeding up the development of existing varieties of wheat, but to making new varieties.

This is of enormous importance for the Soviet Union, and therefore for civilization, today. The Nazis[1] still occupy much of the best wheat-growing areas, and it is doubtful if a full harvest will be reaped this autumn in the reconquered regions.

But vernalization and the breeding of new early varieties have made it possible to grow wheat in northern regions of the Union where summer is very short. This will undoubtedly save many lives, and also spare ships which are needed for munitions. So the investigation of why plants bloom when they do is of immense practical importance.

Potatoes

Lord Woolton is urging us to eat more potatoes instead of bread. Here are what I take to be some of his reasons. In a satisfactory diet we need food with sufficient fuel value, measured in calories, to enable us to work and keep warm. We need enough proteins for growth and repairs. We need lime, iron, and other minerals. And we need small amounts of a number of very different substances which are lumped together as vitamins.

Now if we were to try and live almost entirely on a single food, bread would be the best available. It will give the needed calories, and a rather poor ration of protein, though it needs some supplementing. Whereas potatoes are a grand source of calories, and a

[1] Written in Spring 1944.

fair source of one vitamin, but a rather worse source of protein than wheat. A man could therefore probably live longer on bread alone than potatoes alone. But we are a long way from having to do either, so this does not much matter. Now a good acre of potatoes will yield about twice as many calories per year as an acre of wheat. So, provided we can supplement the spuds with first-rate proteins from meat, milk, or best of all, cheese, a good deal of land is better used for potatoes than wheat.

Of course some land is far better suited for one crop than the other, and the potato crop is much more liable to go wrong than the wheat crop. There are a number of reasons for this. Bread wheats have been grown in Europe for several thousand years, so they have had longer to adapt themselves to our climate than potatoes, which have only been grown for about three hundred and fifty years.

More important is the fact that potatoes are normally reproduced vegetatively. The so-called seed potatoes are not seeds in the ordinary sense of the word, but tubers like those we eat. They are not formed by sexual reproduction, like wheat grains, or the seeds which are formed by potato flowers. Sexual reproduction gives much more chance for variation than sexless reproduction, besides allowing the combination of different qualities by crossing. So it is favourable both to natural and artificial selection, and plants which we propagate from seed, such as wheat, peas, and carrots, can generally be improved more quickly than those propagated by tubers, bulbs, cuttings, or grafts, such as potatoes, apples, and bananas. The potato can be grown from seeds, but none of the varieties breed true. Out of several thousand seedlings only one or two are likely to be better from the farmer's point of view than their parents. And one can only find out whether they are better by testing them in different soils, and finding whether they are immune to a variety of diseases.

Finally the potato belongs to a family, the *Solanaceae*, including nightshade and tobacco, which is particularly liable to virus diseases. These are diseases whose agents are far too small to see with a microscope, though they can be seen with the electron microscope. The viruses of a number of potato diseases can be obtained fairly pure by the same chemical processes which are used for separating different proteins from any mixture of proteins such as egg-white or blood serum. So long as you keep them in a bottle they show no signs of life. But when they are injected into a potato leaf, either artificially through a scratch or naturally by a biting insect, a disease develops. Different viruses cause a number of different diseases, some almost or quite harmless, others fatal.

If, after several weeks, we grind up the leaves, we can extract many thousand times as much virus as we put in, and infect thousands more plants! Now the biologist asks an awkward question. Are we to say that each virus particle is a living thing which multiplies itself? Or shall we say that the plant cells copy it, and produce more of the same kind? The next question is still more awkward. Is there any possible experiment which will decide between the two alternatives? Or have we got to a point where the distinction between living and dead matter breaks down? However that may be, these viruses were worth twenty U-boats to Hitler. Hot weather helps them to spread, so Scottish seed potatoes are generally less infected than English. But you cannot tell an infected potato from an uninfected one without elaborate tests.

A lot of research is being done on potato viruses at Rothamsted and other British stations, but much of it will not be of use for several years. Agricultural research is not only a preparation for peace, but for war. Had Britain devoted as large a fraction of the national income to such research as the Soviet Union, we should have had a good deal less to fear from U-boats.

Britain's Trees

After the war it is proposed to spend large sums of public money on afforestation. We are very short of timber now, and also of soft wood for pulp. That is one reason why we can print so few copies of the *Daily Worker*. As our country has a very dense population, we cannot and should not expect to supply all our needs of timber. But there is no reason why we should not have enough in reserve to keep us supplied for several years if imports are cut off. More accurately there is no scientific reason, but plenty of political and economic ones.

A very large amount of moorland and hillside could be planted, and our woods in the plains could be made far more productive. I have been looking out of the train for the last ten minutes, and though I have seen many trees, the vast majority would be of little use for timber. They had large branches low down, or the trunks were twisted. Often there was a fork where water could collect so that the centre of the tree was bound to rot. They were more picturesque than good timber trees, but much less valuable. The copses were a delightful mixture of trees and shrubs, very well suited to pheasants or other "game" animals, but useless for making tables, pit props, or railway sleepers, all of which we need at present.

The state afforestation schemes will only apply to land which is poor from an agricultural point of view. As long as our present agricultural system goes on, the trees on good land are likely to be neglected. Looking after trees is a skilled job. If a collective farm covered several square miles, one man could look after all the trees on it, except when large planting or felling operations were needed. As long as farms are small, and the interests of the agricultural workers, the tenant farmers, the landlord, and the state, are all different, nothing is likely to be done.

Unfortunately the official afforestation programme does not provide for scientific research. But this is badly needed. No intelligent farmer would think of planting an orchard with cherry trees without specifying the kind of cherry. He would choose his type of cherry, and if he knew a little more, he would choose several different kinds in such a way as to ensure efficient cross pollination. But there are no properly defined varieties of many of our timber trees, and all one can do when planting a new wood is to buy oaks, larches, or whatever one wants, from a good nursery, and hope for the best. One is not likely to get as good trees as one's grandfather could have got. The quality of many of our trees is deteriorating. The reason for this is both good Marxism and good Darwinism.

In a state forest there is a regular routine of felling, and a landlord who spent most of his life on his estate generally knew his trees very well, and preserved the best of them. But a modern landlord is often an absentee, and takes short views about his estate. From time to time he sells trees to a timber merchant, who naturally picks out the best ones. Consequently those which are left to carry on the species are the worst from the point of view of timber production. More and more of our beech trees are heavily branched. On the whole the tendency to branch is inherited, though we do not know the laws of its inheritance in any detail. So the next generation of beech trees, grown from the seeds of those which no timber merchant wanted, will be a good deal worse trees than their parents.

This unconscious selection may improve plants or animals. If one wheat plant gives twice as many seeds as another, it will, on an average, have twice as many progeny next year, so the yield of wheat plants is automatically improved; though because we protect them from competition by weeds, they have become less able to grow in competition than are wild grasses related to them. But if the farmer picked out the best plants in a field to make flour, and left the others to set seed, the yield of wheat would deteriorate.

As our trees have never been selected, consciously or unconsciously, for timber production, it follows that if we started selection, we could probably produce as great improvements in them as our ancestors did in wild wheats, pigs, poultry, and so on. Such improvements have been made in Sweden by the Society for Breeding Forest Trees, a company in which the state holds many of the shares, and which conducts research as well as selling seed. Swedish workers found that a number of valuable qualities in trees were strongly inherited.

They also found something much more remarkable. Among a number of aspen saplings a few grew twice as quickly as the remainder. They were found to be triploids, containing three sets of chromosomes in the nucleus of each cell, instead of two. Triploid plants form very little seed, and what they form is generally sterile. So these plants do not propagate their good qualities by seed, though they can be multiplied by cuttings. If the rapid growth rate were inherited, they would probably have displaced other types of aspen by natural selection. It is, however, possible by a rather roundabout process to breed triploid trees; and Swedish workers have taken the first step in the process of breeding triploid larches, spruces, and firs. This may mean that towards the end of this century most of the forests of Sweden will consist of trees which grow very much quicker than those grown today, so that the yield per acre will be greatly increased.

In this country work on similar lines is being carried out with apples and pears. Here the sterility does not matter; indeed it is rather advantageous, for triploid trees form large fruits with few pips. But nothing is being done with forest trees. It will be no use buying seed from Sweden. For trees are adapted to the climate in which they grow, and not only to the conditions of temperature, but to the length of the day. The Swedish trees do best when the summer days are very long, and though those of southern Sweden might suit Scotland, they would not be at their best in England.

At present in peace time we import large amounts of timber from the Soviet Union. As the U.S.S.R. is industrialized and as its population increases, it will need its own timber. The same will be true of Canada, though here the expansion is not likely to be so quick. In another generation, therefore, Britain may be faced with a shortage of wood in peace as well as in war. So it is not only necessary to plant new forests in our country, but to see that the new trees are properly chosen. If this is not done it will not matter much to me, but some of my readers may live long enough to be almost as short of furniture as they are today.

3

HUMAN PHYSIOLOGY AND EVOLUTION

Temperature

IN a recent number of the *Daily Worker* our Quiz expert asked what was the normal temperature of the human body, and gave 98·6° F. as the answer. A correspondent wrote that it was 98·4° F., and the correction was meekly accepted. Both were wrong.

Either of these figures is fairly close to the average temperature of the mouth during the day, when at rest or doing light work. But the temperature of the interior of the body is a good deal higher. It is easy to measure the temperature of the rectum (the lower bowel) or of the urine, and this is always higher than that of the mouth, and may exceed 100° F. in health. Besides this, some people can swallow a thermometer at the end of a rubber tube, and pull it up again; and a thermo-electric junction between two metals, which is no thicker than a thick needle, can be forced into the leg for some inches with little pain and no danger.

In fact the temperature of the body as a whole is generally well over 99°. Fahrenheit originally chose the human body temperature to fix 100° on his temperature scale, as the measurements of mediaeval English kings were used to define the yard and ell. We now know that definitions of this kind are never accurate, and that weights, measures, and so on are better defined by physical or geographical standards. For example, the Fahrenheit scale is now defined so that water freezes at 32° and boils at 212°; the yard is not defined by the king's arm or step, but by a metal bar, and so on. Even these are not completely accurate, but are quite good enough for practical purposes.

So far from being the normal temperature of the body, 98·4° or 98·6° F. is not even the normal temperature in the mouth. A man whose temperature remained constant would be a freak, like a man whose height or weight stayed steady. For your height is greatest when you wake in the morning, and during the day your spine gradually sags. Whilst your weight goes up suddenly when you eat and drink, and falls suddenly when you excrete, and gradually at other times. I own a balance which is so sensitive that if a man sits in one pan, he rises higher at each swing. It was nearly wrecked by a boy who was so alarmed to see himself losing weight that he jumped out of it and spoiled the knife-edge on which it swings.

The mouth temperature generally varies through one or two degrees F. in the course of twenty-four hours. It is usually highest some time between 10 a.m. and 6 p.m., and lowest between midnight and 6 a.m. when it may fall below 97° F. in deep sleep. In fact 98· 4° or 98· 6° is an average waking temperature. The general average is more like 98· 1°. During steady work it rises to about 99· 5°, and during very violent exercise, such as a boat-race, to well over 100°. In fact there is no such thing as a normal temperature, but rather a normal range of temperatures.

Mammals and birds also have fairly steady temperatures, generally rather higher than in man. The sheep is the hottest blooded of the mammals, so far as we know, with an average rectal temperature of 104°, whilst the tiger appears to be no warmer than ourselves; but perhaps Dr. John Davy, who took the temperature of a tiger in 1839, did not wait till the thermometer had become steady. Birds are much hotter. Many of them have temperatures of 110° F. or over, whereas this temperature is generally fatal to human beings, though a few people have recovered from so high a fever as this.

Other animals have no regular temperature, and are either very little warmer than their surroundings, or some five to ten degrees above it. The one exception to this rule is of great interest. Outside the hive a bee cannot keep its temperature up. But a family of bees in a hive can do so. The temperature in the middle of a hive is usually a bit below that of the human body, but may rise to 100° F. So temperature regulation may be a social as well as an individual matter. In the same way men regulate their temperature by building houses, lighting fires, and so on, as well as by putting on or taking off clothes, and by unconscious processes.

A constant, or nearly constant, temperature is one of the latest products of evolution. The older classes of vertebrates, namely fish, amphibians, and reptiles, do not keep their temperature steady, whereas birds and mammals do so, and have hair or feathers which make it possible. This gives them several advantages. In the first place they can live in cold lands. Even in England reptiles can barely live, and insects die in the autumn or stay motionless through most of the winter. Whereas some tropical birds can live out in a frost if they have enough food to keep them warm.

Secondly, a standardized temperature makes for greater efficiency. Life depends on chemical processes going on in the cells. All of these are speeded up by a rise of temperature, but not all are speeded up equally. So a sudden temperature change throws the machinery of life out of gear, so to say. This is particularly serious in the case of the very delicately balanced processes on which consciousness depends. A small rise of temperature brings on the delirium which is a symptom of high fever. A small fall causes stupor. Other organs, such as the muscles and glands, are much less upset than the brain. Indeed a fairly steady temperature was probably essential before anything but a very rudimentary mind could evolve.

Even within the range of temperatures where thought and skilled work are possible we can see how intimately our minds depend on matter by a simple experiment. An expert can judge the passage of time very accurately without any external aid such as the sun or a clock. If his brain is warmed up by fever, by a hot bath, or by high frequency electric currents, time will seem to him to pass much more slowly than is actually the case. This is so whether he estimates it by tapping (as he thinks) once a second, or subjectively. It is quite easy to halve the rate at which time seems to pass, or in other words to double the rate of passage of subjective time, so that at the end of 30 seconds he says a minute has gone by. The influence of temperature corresponds exactly with that on some chemical reactions, and not, by the way, with the effect on the heart beat, which is also speeded up.

This has an interesting philosophical consequence. Some people think that a disembodied spirit might be outside space (whatever that means) but still aware of the passage of time. But these experiments seem to show that we get our notion of time, as well as space, from the material world. This is of course in agreement with the Marxist view that change is as real and fundamental a property of matter as extension in space, and with the Christian view that the blessed and the damned both have bodies.

The temperature of matter is a measure of the amount of internal change, such as atomic vibration, going on in it. There is always some. No-one can extract all the heat from matter, and bring it to complete rest. And similarly the temperature of any piece of matter is always changing, even though we can keep it much steadier than that of the human body.

So though the notion of a normal temperature is quite useful, any temperature within a fair range is normal. And if we take the notion of normal temperature too seriously we are on the path that leads us to believe in such dangerous abstractions as the Nordic man, the unchangeable laws of human nature, and the instincts of an English gentleman.

Quantity and Quality

Students of Marxism often find the principle of the change of quantity into quality difficult to understand. And opponents of socialism never seem to realize its existence. Some of them say that under socialism no-one would own any private property, even a pair of trousers. Others claim that the Soviet Union is not truly socialist because workers can lend their savings to the state, and draw interest on them.

We can understand the fallacy in such statements if we take some examples from physiology. If these critics were consistent they would go in mortal terror at every breath, because the nitrogen and oxygen of which air consists are deadly poisons—if you have enough of them.

About one-fifth of the air consists of oxygen. We use about half a cubic foot of this gas per hour at rest, and four cubic feet during very hard work. If we breathe any gas, such as nitrogen or hydrogen, which has no oxygen mixed with it, we become unconscious in less than a minute, and die within five minutes. Oxygen is an absolute necessity of human life. Luckily it is so common that nobody has been able to monopolize it.

However, as we go up the air gets thinner. At about 19,000 feet there is only half as much air in a cubic foot as at sea level. Mountaineers can acclimatize themselves to live at this height. But if one goes up to it quickly, as in an aeroplane, one becomes silly at once, and quite ill after a few hours. These symptoms are at once relieved by breathing pure oxygen, or even air to which a fifth of its volume of oxygen has been added. So the crews of aeroplanes need oxygen, and various firms make quite a good thing out of their need.

Oxygen is also used for treating some lung and heart diseases at ground level. But yet it is a poison. Pure oxygen at ground level is not very poisonous, though if one breathes it for two or three days it causes inflammation of the lungs. But at high pressures it is a violent poison. A diver sixty-six feet below the sea is under a pressure of three atmospheres. Before air can be pumped down to him it must be squeezed into one-third of the volume which it occupies at sea level. This is what we are doing, incidentally, when we fill a tyre at 30 lb. per square inch pressure in addition to the 15 lb. pressure of ordinary air. It would be very convenient if we could give the diver pure oxygen to breathe. If so he could come up without waiting, as he would be in no danger from the formation of bubbles of nitrogen in his tissues, which may cause severe pains, called bends, and paralysis.

Behnke and other American scientists have found that at this pressure oxygen affects the brain, and above all, the eyes, so that after three hours a man becomes almost blind. He can only see things straight in front of him, and even then not very clearly. He cannot see sideways at all. Luckily he recovers in a few minutes. When oxygen is breathed at four atmospheres' pressure. Behnke found that convulsions came on after about forty minutes; and very unpleasant they are. Even before this time cramp is said to develop in a muscle which is working hard. At higher pressures oxygen causes convulsions still quicker.

Some organisms are killed by the oxygen of ordinary air. Among them is the bacillus which causes lock-jaw. You will not get lock-jaw from rubbing earth into a scratch. But you may get it if earth containing the spores of the bacillus is carried into a deep wound where oxygen cannot penetrate.

Nitrogen is also poisonous if you get enough of it. If you breathe air at ten atmospheres' pressure, corresponding to a depth of 300 feet, you very soon become rather silly. Divers at this depth often cannot carry out instructions properly, or do skilled work, and the American workers have made it almost sure that this is due to the nitrogen in the air breathed. I have produced further evidence myself.

Finally, water is a poison. If enough water gets into your lungs you are, of course, drowned, but this is not what I mean. You can be poisoned by drinking too much water as surely as by drinking too much beer or whisky. A normal man cannot be poisoned in this way, because he excretes unwanted water with his kidneys. But this excretion can be temporarily prevented by injecting one of the hormones from the pituitary gland; and if this is done, two or three gallons of water will give you convulsions or cramp not unlike those of oxygen poisoning. They can at once be relieved by injecting strong salt solution into a vein, which brings the composition of the blood back to nearly normal.

In the same way everyone knows that you can have too much or too little food, heat, light, and other good things. Aristotle and other Greek philosophers applied the same principle to social occurrences. Aristotle said, for example, that the coward took too few risks, the rash man too many, and the brave man the right amount. And Marx constantly used the principle in his economic arguments. He showed, for example, that a large sum of money could be used as capital, but a small sum could not. No doubt when socialism has developed into communism there will be no such thing as individual savings, for one thing because there will be no need for them. For everyone will get not merely necessities, but many things which we now regard as luxuries, free.

But even the Soviet Union is still a long way from communism.

And under communism there will doubtless be some private property. If we understand how quantity is transformed into quality, we shall realize that private property, like oxygen, can be both a necessity, as in the case of boots, and a public danger, as in the case of armament shares. And we shall steer our way between the extremists of the left, who think that a Soviet worker is a capitalist because he lends a few hundred roubles to the state, and those of the right, who think that because I can own a fountain pen, the Duke of Westminster should be allowed to own hundreds of acres of London.

Blood

As the destructive power of modern weapons increases, so do the resources available for treating the wounded. Probably the commonest of all results of a wound is loss of blood. This may be fatal by itself. Or it may lower the resistance so that the wounded man or woman succumbs to an injury which they would otherwise survive. Fortunately loss of blood is the easiest of all major injuries to remedy.

This is exactly contrary to the primitive ideas of physiology which are still current, especially in advertisements and sermons. Primitive men sometimes identified life with blood and sometimes with breath, for the reason that people die quickly if they lose blood or stop breathing. This idea is perpetuated in the word "spirit," which originally meant breath, and in the Nazi idea of blood as something peculiar to a race. We are also misled by advertisements which attribute all sorts of diseases to impurities in the blood. Actually the blood is simply the transport system of the body, and is full of waste products on the way to be got rid of by the lungs or kidneys. But very few diseases are due to their accumulation, and they may rise well above the normal level without any serious loss of health.

The blood is perhaps the least living part of the whole body, and once doctors began to give up these ancient ideas they found that it was the easiest to replace. At present loss of blood is dealt with in two distinct ways, by injecting whole blood, or plasma. About half the blood consists of a clear slightly yellow fluid called plasma, which contains water, salts, sugar, and proteins. The other half consists of corpuscles, mainly red ones concerned in carrying round oxygen from the lungs to the various organs where it is used. An average man has a volume of blood equal to about one-twentieth of his bulk, that is to say less than a gallon in most people.

One can lose a tenth of it without noticing any difference; after losing a quarter one feels fairly faint; and after losing half one is likely to die. This is not due to the loss of corpuscles. An anaemic man or woman with only half the normal number of corpuscles can still work, though he or she is rather weak and easily fatigued. But if the total volume is reduced, although the blood vessels contract, the heart is unable to pump the blood up to the head, and in consequence it is advisable to lay the patient flat and lift up the legs to let the blood drain out of them and give the head as good a supply as possible.

An obvious idea would be to inject pure water to restore the volume. But this would be very rapidly fatal, as the water runs out of the blood into the body cells which swell up and burst. Much better results are obtained if we inject water with the same amount of common salt as is found in blood.

But this is not enough either. The correct salts to add were discovered by Professor Ringer at University College, London, in a rather odd way. He was studying the effects of drugs on frogs' hearts placed in salt solution. During one season he found that they were beating much better than ever before. After careful detective work he found the reason. His laboratory assistant was no longer making up the solution with distilled water, as he was told to, but with London tap water, which contains a fair amount of lime from the chalk hills round London. Ringer found that hearts would beat for a long time if perfused with water containing a little calcium chloride and potassium chloride as well as sodium chloride. Later workers found that the blood of primitive sea animals is of nearly the composition of sea water, while fresh water and land animals, and also most fish, have a blood plasma like diluted sea water, and probably like that of the sea several hundred million years ago.

If Ringer's solution of salts is injected into an animal which has lost much blood, it recovers strength for a short time, but the added water and salts leak out in an hour or so. They will stay in the blood vessels if the proteins of plasma are added also. So a good deal of the blood which is taken from donors is now put in a centrifuge and spun. The fluid part, or plasma, is then dried, and will last for many months. It can be diluted with about 15 times its volume of water, and injected. It is far more portable than whole blood, and lasts longer. So it is particularly useful on ships, and in mobile dressing stations. But a wounded man whose blood volume has been brought back to normal with plasma is still short of red corpuscles like a sufferer from anaemia, and may take a month or so to make all the new corpuscles that he needs.

Whole blood contains corpuscles as well as plasma, and this advantage brings a danger with it. Human beings fall into four groups as regards their corpuscles, and in some cases corpuscles from a member of a different group are destroyed when injected, and do more harm than good. Fortunately the corpuscles from one of the groups, called group O, can be injected into almost anyone with safety. The only known exceptions, and they are rare, are found among pregnant women and those who have recently borne a child. All these facts took a long time to discover and the discoveries were made all over the world. Thus the basic ideas of plasma transfusion came from Ringer and Bayliss at University College, but the modern technique was worked out in America. The blood groups were discovered by Landsteiner in Austria and Jannsky in Czechoslovakia. The first work on blood storage was done by Rous in New York, and the modern methods are largely due to Briukhonenko and others in the Soviet Union.

Surgeons are now trying similar experiments with other tissues besides blood. A good many blind people have had their sight restored by grafting bits of the cornea (the transparent window in front of the eye) from a healthy eye on to a diseased or injured one. Here again the original technique came from Austria, but was greatly improved when Filatov of Odessa found that a cornea could be grafted from a dead eye. These techniques sound simple, but only a physiologist realizes how many years of work, much of it apparently useless, have been needed to perfect them, and how completely this work has depended on international co-operation.

Blood Analysis

Any day now your ear may be pricked, and a sample of blood taken from it to estimate the haemoglobin. An official inquiry is being made into the nation's blood, or rather of one particular substance in it, and samples are being investigated from a number of different age and social groups.

The blood probably carries round thousands of different substances. It carries oxygen from the lungs to the organs which need it, and also food and water from the intestines and from depots such as the fat under the skin. It also carries away waste products to be removed by the lungs, kidneys, and skin. Finally, it carries round hormones from one organ to another. If it did not carry the hormones from the pituitary gland children would stop growing and adults would sink into a state of lethargy. If it did not carry the hormones from the ovary or testicles, men and women would lose many of the characters which distinguish them.

These various substances have not merely to be in the blood, but to be there in the right amounts. With too little oxygen you become weak and silly, with too much you have a convulsion, in each case losing consciousness. I have tried both, and prefer too little. Too much urea is a sign of kidney disease, too much sugar of diabetes. With too little thyroid hormone you become a fat, sluggish imbecile; with too much a thin, jumpy neurotic with a ravenous appetite. Some of these substances are easy to estimate chemically, but about a teaspoon full of blood is needed in most cases. Others, including the hormones, are much harder to determine.

The easiest of all is haemoglobin, the red substance in the corpuscles which carries oxygen. More accurately it is red when combined with oxygen, purple when uncombined. In 1900 my late father produced the first quick and accurate method for estimating haemoglobin from a single drop of blood. He diluted a measured amount with water, added coal gas to form the compound of haemoglobin with carbon monoxide, and then added water drop by drop till the colour matched a standard solution. The more haemoglobin in the blood, the more water had to be added.

He analysed the blood of a number of men, women, and children. I was one of the children, but it was rather hard to get all that he wanted, and my sister was heard at our door telling a young friend "You come in here, my farver wants your blood." There was a good deal of difference between apparently healthy individuals, but women and children had, on an average, a good deal less haemoglobin than men. For over thirty years this was accepted as a natural phenomenon, like the lesser average height of women.

Then McCance found that, even among the well-to-do classes, women were generally short of iron, and that if they were given enough of the right iron salts the amount of haemoglobin in their blood went up. They need more iron than men, as they normally lose blood each month. So another alleged inborn inferiority of the female sex went west. On an adequate diet women may make as much haemoglobin as men. But in peace time most women were not getting enough iron, even when they could afford it.

In war time it is hard to get some of the foods which are the best sources of iron, such as liver, winkles, and chocolate, though cocoa is not hard to come by. And a shortage of iron is not the only food shortage which causes anaemia. The red corpuscles only last about six weeks on an average, and new ones are constantly being made. A shortage of proteins in the diet, or proteins of the wrong sort, can slow down the production of new corpuscles.

The research which is now under way will show up any dietary deficiencies which interfere with the manufacture of new blood. If any groups of the population are getting anaemic—which is not certain—the most likely are housewives, and children who are growing very quickly. There can be no question of serious malnutrition for those who eat at factory canteens or British restaurants. And on the whole there is probably less malnutrition than before the war, because the nation's food is being shared out more fairly than ever before. However, the Government is quite right to look for evidence for it.

If such evidence is found, and a slight redistribution of our present supply is needed, let us hope that the cuts will be made where they will do least harm, in the supply of food to expensive restaurants. But meanwhile, the readier everyone is to provide a drop of blood, the quicker the information will be available.

Blood and Individuality

Early this year a baby was lost in one English town, and a few weeks later a baby of the same sex and roughly the same appearance was found in another town. Blood tests are now being done to help to find out whether these two are the same baby.

We know a lot more about the chemistry of the blood than that of any other human tissue, because it is the only one which can be taken out in quantity without harm. Muscles, brain, and so on, can only be obtained after death when they have changed a good deal, or in an unhealthy condition at an operation.

The chemical analysis of blood is an extremely skilled job. It took me about three months before I could get duplicate analyses of the amounts of oxygen and carbon dioxide in a cubic centimetre of my blood to agree within one per cent or so. Chemical analysis is valuable in detecting some diseases, and finding out accurately what happens in others. For example, if you find three parts of sugar per thousand, instead of about one, it is fairly sure that the blood comes from a severe case of diabetes. If the phosphate in a growing child is down to the adult level, the child cannot make new bone, and is in danger of rickets. And so on.

Besides substances of accurately known composition, we can detect others whose make-up is only roughly known. Thus the blood from a patient who has, or has recently had, typhoid fever, contains proteins which combine with typhoid bacilli, and make them clump together. This is a useful test for typhoid. But all the substances mentioned so far may alter in quantity in the course of a week, or even of a few minutes. They could not possibly be used for identifying a baby, unless it were suffering from a chronic disease which altered the blood's composition.

There are, however, some substances in the blood which differ from one individual to another, and never appear if they were not there at birth, or disappear if they were so. The most important in practice are those which determine membership of a blood group. Everyone belongs to one or another of four groups, and everyone should know to which group they belong, for a very simple reason. After an accident someone may need half a pint of your blood, or you may need half a pint of someone else's. Most doctors under 35 can probably determine blood group membership as well as I, which means that they might make one mistake in a hundred. But an expert at a blood transfusion centre makes mistakes at the same sort of rate as a good railway signalman, less than once in a lifetime. No doubt I could get into this class with a few weeks' practice; but I have other things to do, and so have most doctors. So if you have once been grouped by an expert, you may save your own life or someone else's by protesting if a non-expert assigns you to the wrong group.

Membership of the blood groups is hereditary, and the rules according to which membership is determined are well known. Exceptions to them, which occur with a frequency of about one per thousand in some communities, and a good deal more in others, can almost all be explained by illegitimacy. The simplest rule is that no child can have an antigen on its red corpuscles which is not present in one parent or the other. If, for example, neither of a baby's parents have the A or B substance, the baby cannot have them. In other words, if the parents both belong to group O, and the baby does not, it is no child of theirs. Besides the characteristics which determine blood group membership, the blood corpuscles show a number of others which are of no practical importance, so far as is known, but whose inheritance is understood, and which are constant throughout life.

It will not be possible to say that this baby is certainly the child of its alleged parents. But one of the very few men and women in this country who combine the necessary technical skill and knowledge of heredity should be able to say one of two things. Either the baby is not the child of the couple who hope they have found their lost child; or it has a combination of characters which would be found in only one child in ten or twenty picked at random, but could be found in one of their children. If so the parents will probably accept the baby.

The main practical use of such tests has been in cases of disputed paternity. Here too one can never prove paternity for certain, but one can disprove it. The most important use will probably be to work out the origins of the human peoples. So far they have been classified on their skin colour and the shape of their hairs and skulls. But skin colour at any rate is not only variable in the individual, since we get browner in summer, but is an adaptive character in a race. Many tropical peoples have black skins. This is probably an advantage, in protecting them from sunburn. So it may have been developed independently in different areas, and there is no reason to think that, for example, Australian blacks are any more nearly related to African blacks than to Europeans.

But the blood group characters do not alter during life, and are of no particular use to those who bear them. They show that there is no such thing as a pure race. Every people contains members of groups O and A, and mostly of all four groups, even if their skins and hair are very uniform. But the proportions differ greatly, and often in an unexpected way. Thus the Irish differ far more from the English than do the French or Dutch.

The case of this baby brings out one fundamental point. There are human characters which are determined entirely by heredity and not environment. But these are seldom of any biological or social importance, except where there is a prejudice against a skin colour or nose shape associated with a particular race. The characters which matter most, whether they relate to physical strength, or to mental or moral ability, are almost always influenced by environment, and can be improved by improving society.

How Muscles Work

Self-movement is one of the most obvious properties of animals, including men. At first it was regarded as a property of living things alone. Then men made machines which moved themselves, and in the seventeeth century philosophers began to explain animals as machines. Descartes thought that a muscle shortened because it was blown out by "animal spirits" pumped down the nerves from the brain.

This is wrong, for the nerves are not tubes, the muscle can be made to work by very mild electric shocks after the nerves are cut, and its volume slightly decreases when it shortens. In the nineteenth century it was found that a man and a machine produced just the same amount of work plus heat when a given amount of food was combined with oxygen, whether it was burned in a furnace or oxidized more slowly in the body. Later it was shown that most of this oxidation occurred in the muscles themselves, the main substances used being sugar and oxygen from the blood. This was very important, as it is very important to know that a power station uses so much coal, water, and lubricating oil per kilowatt-hour; but it told us little of what went on in the muscle.

The first information on this matter came from the fact that a muscle can work for some time without oxygen, though it loses efficiency for a while as a result. You can see at once that this agrees with the common sense of athletics. The long-distance runner is limited by the rate at which his lungs can supply his muscles with oxygen. He cannot keep up a steady speed greater than corresponds to this. The sprinter works his muscles much faster than their oxygen supply warrants. When he stops he absorbs extra oxygen for some time. If the long-distance runner sprinted at the start he would have to run the rest of the race with his muscles short of oxygen, and would lose a lot of speed.

There is only a very small store of oxygen in the muscles. Biochemists gradually found that various substances accumulated in muscles which worked without oxygen, the first of these being lactic acid, isolated by Fletcher and Hopkins. During the recovery process, when oxygen is used, these substances are put together again. A muscle may be compared to a submarine, which uses oxygen from the air for its motors when running on the surface, but can run for a long time on its accumulators below water without using any oxygen. On reaching the surface the accumulators are recharged. But the muscle almost certainly differs from the submarine in one respect. The source of energy for contraction is always the process not needing oxygen, corresponding to running the ship off the accumulators.

The next phase of research was to trace the various chemical changes occurring in a muscle both during contraction and recovery. A large number of hitherto unknown chemical compounds were isolated, and a number of enzymes were found which cause them to change as they actually do. That is to say, by adding a particular extract of muscle to a certain pair of substances which did not interact without it, they would form new substances. The enzymes which make the reactions go forward are all proteins, and probably most of the proteins in muscles which we digest when we eat meat, are enzymes.

Meanwhile biophysicists examined muscles, both relaxed and contracted, with X-rays, and discovered a good deal about the protein called myosin which forms the microscopic fibres whose contraction shortens the muscle. The long molecules become crinkled, as a chain of steel links might do if each were magnetized so as to attract its neighbour.

In 1939 Engelhart and Lyubimova in Moscow made a very fundamental discovery. Myosin, the protein which contracts, is also the enzyme responsible for a particular chemical change which gives rise to a good deal of heat in ordinary laboratory experiments. But in the living muscle the energy is not wasted as heat, but much of it is used as work. Bailey and Needham confirmed their discovery in Cambridge. The biochemists had been thinking of proteins as enzymes, the biophysicists as part of the contractile mechanism. The Soviet workers thought of them more dialectically. The same protein changes adenosin-triphosphoric acid, and, in doing so, changes itself. If it isn't alive, it is getting near being so.

We are still a long way from making an artificial muscle, but

D

we have at least an idea of how it would be made. It would be more efficient than any engine of its size, that is to say it would give more work for a given amount of fuel. But it would produce much less power per unit weight of working parts than many existing engines. It would probably be a bit lighter than a living muscle of the same power, because, not being alive, it would not have to grow or repair itself, but it would be no use for an aeroplane, probably none for a motor vehicle. However, as a source of power in a stationary engine it might well be really useful.

Meanwhile the main practical result of Engelhart and Lyubimova's work is likely to arise when similar work is done for the heart, which works rather differently from most muscles, and goes seriously wrong much more often. Unfortunately work of this kind was almost wholly stopped during the war, and is not restarting very rapidly in England. For example, the physiological laboratory at University College, London, is still occupied by the Admiralty. The liberation of such laboratories is indispensable before we can attack heart disease on these lines.

Sense Organs

In one of the air raids of this winter the *Daily Worker's* office and plant was damaged.[1] It did not receive a direct hit, but a bomb bursting near to it broke many of its windows. And a lump of concrete from a neighbouring building set off the system of sprinklers designed to put out fires. The water damaged a lot of paper and some machinery, and brought down a good deal of plaster.

This accident interested me, not merely as chairman of the editorial board, but as a biologist. For it was an example of a response to the wrong stimulus. The sprinklers are supposed to act in the event of fire, but on no other occasion. In the same way a sense organ such as the eye, if it were perfect, would give sensations of light only when light entered it, and not when it was struck. But a thousand million years of evolution have not made the eye into a perfect organ. In fact in some respects, though not all, it is less perfect than the photographic camera, which is

[1] The office was burned out and much of the plant destroyed soon after its suppression.

only a century old, but has been deliberately designed, whereas the eye seems to have evolved through unconscious struggle.

To show how sensitive the eye is to slight pressure, press your eye gently with a finger near the bridge of the nose, through the eyelid. You will see something, usually a bright ring with a dark centre. If you press the inside of your right eye, it will appear far out to the right. If you move your finger up, it will go down. The sensitive film at the back of your eye, called the retina, is affected by pressure as well as light, and much more so than a photographic film. Since the images thrown on it by the cornea or front transparent window and the lens of the eye, are upside down, and rightside left, as in a camera, it is natural that the bright ring appears where it does.

You can also stimulate the eye, or any other sense organ, with an electric current, though I do not advise you to try. Still less do I advise you to try a chemical stimulus. But you can try this on another set of organs. Your skin and the inside of your mouth are full of microscopic sense organs giving you feelings of warmth. On your hand or face these are pretty close together, but elsewhere, for example on the front of the thigh, they are so far apart that they can easily be mapped by finding where a warm piece of glass or metal is felt to be warm. These organs are easily excited by mustard and several other substances. In fact right up to the time when thermometers were invented, philosophers discussed whether mustard was really hot. Nowadays it is pretty universally admitted that the thermometer is a better indicator of heat than the skin, and that the heat of mustard is an illusion like the sparks which we see when the eye is hit.

For efficiency, not only must each organ respond to one particular kind of stimulus, such as light, sound, or heat, but it must be connected up with the nervous system, and through it with the muscles and glands, so that when it is stimulated an animal or man responds in the right way.

Simple animals respond in very simple ways. A clam shuts its shell if the light is suddenly diminished, or if it is touched or shaken, and thus protects itself. Other animals move towards light or darkness, or away from strong stimulation of any kind. Most animals have some sense like our taste and smell, by which they recognize food; but to recognize anything by sight, sound, or touch, an animal must react to a pattern.

The pattern may be in space, like the shape of a man or a bicycle, or in time, like the sound of a melody or a word. But it must be recognized against a number of different backgrounds, and under slightly different forms. The same object, seen in different directions, and at different distances, makes a different pattern on the retina. And a brain is needed before an animal can react in the same way to these different patterns. Some bottom-living fish will eat a crab on the bottom of the tank where they live, but will not touch one hung by a string on a level with their eyes. They cannot recognize "crab in the water" as the same as "crab on the ground." We spend much of our first year in learning to recognize simple things.

Although I am not one of those who think that the brain is a machine, it is probably acting in a mechanical way when it recognizes a simple shape or melody. For machines can be constructed which will open a door only when a particular melody is played or a particular word pronounced. Whether a machine could be made which answered to the same word pronounced by Willie Gallacher, Harry Pollitt, Isabel Brown, and myself, is more doubtful.

Philosophers have long disputed as to what there was in common between all squares, or all dogs. The reactionary Greek philosopher Plato said that the squares were imperfect copies of the eternal idea of squareness, and the dogs of dogginess. He used this theory to attack democracy. For he thought that all states should be as near as possible copies of an ideal city, which was far from democratic. His disciples today oppose the notion of indefinite progress with the notion of eternal values. I think that we shall know a lot more about what there is in common between all squareness, or all dogs, when we understand how the brain analyses our sensations to pick out the common factor.

To go back to the sprinklers in the *Daily Worker* office, their "sense organ" is a glass bulb containing some fluid which boils easily. When the temperature rises beyond a certain point the bulb bursts. This releases a plug in a pipe containing water under high pressure, and a jet of water spurts out. Obviously such a device will break with a violent shock as well as with heat. The same applies to some other heat-detecting mechanisms. In this case the appropriate response is a spray of water at each point which is heated up. This is a local response like the mechanism on a sea-anemone or jelly-fish which stings its prey at any point on the tentacles that is touched. It does not need anything like a nervous system. In some large buildings a fire rings a bell in a particular room, and switches on a lamp which shows where the fire is located. For a fuller analogy to a nervous system we should have to imagine a small fire engine which automatically ran to the place of the fire, guided by signals from the control room.

At the present time our instruments are much more sensitive than our sense organs. The telescope photographs stars too faint to be seen, the balance weighs objects too small to be felt. Other instruments can pick out patterns which we cannot detect. For example, a tide-predicting machine picks out periods which the unaided eye and brain miss. As machines of this kind are developed we shall learn new facts about nature, just as we did when the microscope was invented.

The remote control of machinery, corresponding to the motor side of our nervous system, is being studied, particularly in the Soviet Union. As progress is made along these various lines our descendants will doubtless learn to know and control aspects of nature which today we can no more imagine than our ancestors five centuries ago could imagine electric currents. And by studying these artificial senses, brains, and muscles we shall not only increase our control and understanding of nature, but of ourselves. Without such understanding we react as automatically and irrationally as the sprinklers in the *Daily Worker* office.

Brain Waves

In a recent murder trial it was shown that the criminal's brain produced abnormal electric waves, and on the strength of this the jury sent him to Broadmoor instead of to the gallows. This was supposed to be an act of mercy. I am not so sure. I am going to die anyway, but I hope I shall not be imprisoned for life. In such a case hanging may be the more merciful treatment.

The electrical oscillations which saved the man's life are among the many biological facts which were discovered when modern radio technique was applied to physiology. All the organs of the body produce electrical changes when active, but the potentials produced are measured in thousandths of a volt at most.

The easiest of all to detect is that produced by the heart, and this was the only one which had been recorded adequately without the use of an amplifier. You put your hands, or a hand and a foot, in bowls of salt water with a wire leading from each; and these wires are connected by a fine quartz thread, silvered to make it conduct, and passing between the poles of a magnet. At each heart-beat a small current passes along this thread, and its movements can be photographed. The current can also be amplified and recorded with an oscillograph using a beam of cathode rays like that in a television set, which is deflected at each heart-beat. The record is called an electrocardiogram in either case. Or it can be made into sound by a loud speaker. The electrocardiograms differ according to where the two electrodes are placed, but their most striking differences are due to heart disease. When my heart started occasionally missing a beat I knew that this might mean nothing at all, or something very serious, and was greatly relieved when the electrocardiogram unmistakably showed the former.

If you place your electrode against a nerve, and put in a good amplifier, you can overhear the messages going up and down it. If a needle is placed against a motor nerve controlling a muscle in a man's arm, and the electrical changes amplified, one normally hears an occasional crackle, for a muscle is never absolutely flabby. But if he contracts the muscle, say by clenching his fist, one hears a roar like a dozen machine guns, as hundreds of messages pass down per second to make the muscle contract.

The electrical changes in the sense organs still go on in an anaesthetized animal, though the receiving stations in its brain are out of action. So the messages from the eye to the brain can be tapped and partially decoded without causing pain. In this way the Swedish physiologist Granit has just shown that the sensitive film at the back of the eye, the retina, behaves like certain types of plate used in colour photography. Some of the cells in it are highly sensitive to red light, others to green, and others to blue, and we judge of a mixed colour on democratic principles, according to the numbers of the different types sending messages to the brain.

Finally one can take records from the brain itself. Naturally you get the best results if you saw through the bone and put one electrode on the brain itself. This has now been done on hundreds of human beings during brain operations. Many of them were fully conscious, a local anaesthetic being used where necessary. No pain is felt when the surface of the brain itself is stimulated electrically, or even cut; for the parts concerned with pain are deep down in the middle. On the other hand the surface is certainly engaged in feeling, willing, and thinking. So one can actually take records of the activity of the cells concerned in consciousness.

The greatest electrical activity in the human brain comes from the part at the back which is responsible for vision, to which nerve fibres run from the eye. This produces enough effects to be detected even through the skull, so no operation is necessary to record them. If the eyes are shut, the main feature of the electroencephalogram, as the record is called, is a series of waves with a rather irregular period of about a tenth of a second. These may persist if the eyes are opened, especially in dim light, but if the subject looks at anything attentively they at once break up into an irregular disturbance. They do the same if he imagines anything visually, or does mental arithmetic or other thought processes which involve visual imagery. The large waves are caused by millions of cells discharging nearly simultaneously. But if they are doing different things, some being concerned in seeing white, others black, the unison is broken up.

The technique of recording these waves was invented by Berger in Germany, improved by Adrian in England, and applied to medicine by Gibbs and his colleagues in America.

Epilepsy is, unfortunately, fairly common. An epileptic may have violent convulsive fits, or he or she may merely lose consciousness for a short time, without falling down or making any violent movements. Between fits epileptics may seem quite normal. But the rhythm of the electrical brain waves is slowed down or speeded up according to the type of epilepsy, and the condition can thus be detected in between fits. Epileptics do not know what they do during a fit, but they are conscious at other times, and if an epileptic commits a crime, he may have known quite well what he was doing.

The scientific and philosophical interest of these records of brain activity is, of course, very great. But no-one has gone very far with their interpretation. We can no more guess from these records what a man is seeing or imagining than we can tell from a rather bad gramophone record of a factory whether it was making guns or tanks. But it took over half a century before records from nerves were got clear enough to distinguish between impulses carrying messages of pain and of other sensations. There can be little doubt that as technique is improved the electro-encephalograph will help to clear up the nature of the processes going on in the brain when we feel, think, and will.

Learning

A comrade in the Corps of Signals has sent me a question which I cannot answer. But it is so interesting that I am going to devote an article to it. He wants to know what happens when a man learns Morse, finding it difficult at first, but finally sending signals or taking them down automatically.

No doubt some change takes place in his nervous system. We can suggest possibilities, but that is all. The human nervous system consists of a great mass of cells in the brain and spinal cord, numbering several thousand million, or more than the whole human race, and of fibres which extend from these cells, and which die and cease to conduct if their contact with the cells is severed. Some of these fibres take messages in to the nervous system, others take them out. The majority run from one cell to another. The ingoing or afferent fibres include those from sense organs such as the eye and the organs of touch in the skin, and others whose messages do not give rise to sensation, but to reflex actions. For example, we usually regulate our breathing and heart-beat without knowing anything about it. The outgoing or efferent fibres include motor fibres to muscles, which cause them to contract, and inhibitory fibres to some muscles such as the heart, which goes on beating when the nerves to it are cut, and which may therefore have to be slowed down as well as speeded up. They also include fibres whose messages increase or diminish the secretion of glands, for example, sweating or the production of gastric juice.

We can detect the messages going down a single nerve fibre electrically, and amplify them so that they can be heard on a loud speaker. We can measure their speeds, the heat which they generate, and so on. They consist of single impulses like Morse dots, and their effects depend merely on the connexions of the nerve fibres, and the numbers of impulses arriving down a fibre per second. They are simply waves of chemical change moving along a nerve, and not particularly mysterious.

The simplest reactions involving the nervous system are called unconditioned reflexes. Some, such as the contraction of the pupil when a light is shone into the eye, are quite out of the control of the will, except in a few rare individuals. Others, such as the withdrawal of a limb when it is pricked, are reflexes, although the same muscles can also be controlled by the will. Some of these reflexes can still be carried out by the spinal cord after this has been severed from the brain by a wound, though there is no consciousness of pain or movement. But they can be controlled and overridden by the will to some extent in a normal man. Even the simplest reflex involves a number of nerve cells in series. And most of the time between a stimulus, such as a flash of light or a prick, and the resulting muscular contraction, is occupied, not in passage along fibres, but in passing the message from one cell to another.

These reflexes are inborn. Some are present at birth, and some develop later. But they are not learnt. Besides the inborn reflexes there are several kinds of conditioned reflexes, which depend on experience. The most thoroughly studied are those of glands such as the salivary glands, which are not controllable by the will. A hungry man begins to secrete more saliva not merely when he tastes food, or even sees or smells it, but when you talk to him about it, or ring the dinner bell. Pavlov's work on dogs was mainly based on the conditional secretion of saliva. A dog was fed immediately after sounding one note, but not after sounding another, and salivated on the food signal only. Some dogs could distinguish notes only a semitone apart. These experiments were made with a great variety of stimuli. Mr. Bernard Shaw thinks they were cruel. I wish he had seen, as I have, a dog pulling a laboratory assistant along a corridor in a Leningrad laboratory in his eagerness to get to the experimental room. I confess that that if I were their subject I should regard such experiments as rather dull. This dog apparently didn't.

Other conditioned reflexes, certainly in man, and probably in animals, are carried out with muscles controlled by the will, and involve will as well as consciousness in their development. Such are those involved in any skilled activity, such as learning to cycle or transmit and receive Morse. Consciousness is only biologically useful where a choice is involved. We walk unconsciously until we have to decide which side of an obstacle to take or till something else interferes with a series of reflexes. Then memories help us to decide what to do, and large numbers of brain cells become active. But when the same choice has often been made in similar circumstances, for example tapping — ·· for D, consciousness is short-circuited, so that we can think of something else while carrying out the reflex.

An essential feature of learning a skilled activity is that a new pathway for nervous impulses is laid down in the brain, involving fewer nerve cells than before, and therefore quicker response and less interference by other nervous activities. The new paths are not laid down by the growth of new fibres, but probably by increasing the sensitivity of nerve cells to particular forms of stimulation. We know, for example, that some nerve cells will not respond to a single impulse along a fibre. They need a series, neither too slow nor too fast, or simultaneous impulses from several other cells.

But the technique for studying the conditions needed to excite a single cell is so difficult and so new that no-one has yet been able to study changes in its sensitivity. Those who could best do so are at present working either on such problems as the healing of severed nerves, or on radio-location. We do at least know, from a study of injuries to the human brain, whereabouts the paths concerned in learning Morse are situated. They are largely on the left hand side of the brain, above the ear. In so far as sight is involved they extend backwards also. An injury to the front part of the brain does not affect manual skill, though it may lower initiative or affect social behaviour adversely.

If I cannot answer my correspondent, I can at least tell him that the answer to his question will need the full use of electrical apparatus which he probably understands better than I, such as amplifiers and oscillographs, to record the electrical activities in nerve cell and fibres without injuring them. If after the war even a tenth part of the skill which is now used in the Royal Corps of Signals is employed on the recording and decoding of nervous impulses, and on producing them experimentally, his question will be answered in ten years' time.

Fatigue during Skilled Work

Although this war has undoubtedly slowed down scientific research, a great deal of interesting work which is now secret will be published when it is over. Among this, I hope, will be much of the work which has been done on the physiology and psychology of men in the fighting forces.

In a lecture to the Royal Society given in 1941 but only just published, Professor Bartlett describes work done for the Flying Personnel Research Committee on a number of highly trained men. They sat in front of a panel containing a large number of signals for action, including a speed indicator and a direction indicator which had to be watched all the time, others which were supposed to show the amount of petrol available, and so on. The indicators were pointers on dials, or coloured lights. Sometimes the subject's chair was also tilted as it would have been during work. When I add that the operator had to work controls with his hands and feet in answer to his various sensations, especially of the movements of the dials, I think few readers will fail to guess what the experimental set-up represented, though this is nowhere stated.

Professor Bartlett investigated the onset of fatigue in his subjects. Now fatigue of a great many processes has been studied. The earliest scientific studies were of fatigue in hard work such as weight lifting or running. Later workers investigated fatigue of less mechanical but still monotonous work such as copying or adding. But very little had been done on the highest forms of skilled work, such as are needed in industry, transport, and war. And fatigue in skilled work is something very different from fatigue of the muscles. To begin with, the weight lifter does less work in his last hour than his first, apart from a possible spurt at the end. Bartlett's subjects often did more work at the end than the start, because they were making needless movements which they could no longer control.

The most striking symptom of fatigue in skilled work was a lowering of the standards of performance. Even at the end of two hours, errors in direction increased five times, and errors in speed were doubled. Yet most of the subjects thought they were doing better at the end than the start. The mistakes were not due to mistaken actions, such as pulling the wrong lever, or pulling the right lever the wrong way. They were mainly due to wrong timing. Another source of error was a failure to respond to all the instruments on the control board. At first, for example, the subject never neglected the petrol gauge. Later on, he frequently "ran out of petrol," and towards the end he could only attend to one instrument at a time. If this happened he frequently "pulled his machine to disaster," fortunately only on paper.

Early in the experiments sensations due to tilting or vibrating the chair were a help. They allowed the operator to keep a steadier course. Later on they could not be properly interpreted, and merely distracted him. The same is probably true in industrial work. Early in the day of a worker looking after a number of machines, small changes in the sounds they make are a help. Later on the noises are merely fatiguing.

The subjects' own account of these experiments were most interesting. They put down their mistakes to faults in the machinery, saying that the levers were becoming sticky. They not merely failed to notice things that happened. They reported things which never happened at all such as loud noises. They became uncomfortable. They said they were too hot or too cold, that their equipment was too heavy or too tight. They often got cramp. Finally they lost their tempers. At first they worked silently, then they sighed; later a few bad words emerged. "By the end of the experiment," writes Professor Bartlett, "most operators kept up a flow of the most violent language they knew." Such observations have an obvious bearing on complaints of rudeness or lying by fatigued factory workers.

Doubtless these experiments have been extended to deal with the effects on fatigue of oxygen want, noise, cold, and darkness, the proper design of instrument panels, controls, and so on, and the need for rest periods during and after prolonged operations.

After the war it is essential that similar work should be done with skilled industrial workers. Industrial "psychologists," at least in capitalist countries, have been mainly concerned with monotonous manual work, and not with the much higher form of skill needed in looking after a complicated machine, or set of machines. A good many of Professor Bartlett's findings are only commonsense. The important points were to find out how soon fatigue became dangerous, and how to combat it; in fact not merely to describe it, but to measure and prevent its onset.

The Labour Movement should see to it that after the war the men who have worked on fatigue in the Fighting Services are used for the problems of peace time. A tired lorry driver can be as dangerous as a tired pilot. This danger would be reduced by shorter hours and greater comfort.

The study of industrial fatigue could and should benefit enormously from accurate studies of this kind. But unless the Trade Unions take a hand in the matter they will be used to boost profits rather than to promote the safety and health of the workers.

Hygiene or Sales Talk?

A reader has drawn my attention to an article in the current number of *Everywoman*, a magazine which is obviously read by many working-class women. The article deals with cleanliness, and if every woman followed its advice, the national consumption of soap and cosmetics would certainly go up vastly. This would hardly help to win the war.

I am all for as much cleanliness as óne can manage in peace-time. I used to take a daily bath, and a good full one too. But I realized that this was a luxury, and in no way necessary for health, so I have cut it out. The traditional weekly bath night is all you need unless you are in a dirty trade. It is desirable to have a bath and change your underclothes about once a week. Other-wise if you pick up a louse, as anyone may do, you give it a chance to have a family before it is drowned.

One theme runs through the article, namely that it is better to smell like a chemist's shop than a human being. "Haven't you noticed the heavy odour which even the most well-bred scalps develop a few days after a shampoo?" "Happily there are few women today who don't use a deodorant." "To be sure of perfect freshness, you should wash every part of your body both night and morning." And so on. Human beings have a natural smell, which a large fraction of others find attractive; and still more would do so but for sales talk about body odour. Of course, if you leave any part of you unwashed for too long, bacteria get to work and produce a less pleasant smell.

Your skin contains millions of sebaceous glands, mostly near hair roots, which secrete grease. This grease is part of your normal equipment. It probably protects your skin from bacterial invasion, and keeps it soft. Soap removes it, though a bath with-out much soaping leaves a good deal. I suspect that quite a lot of minor skin trouble comes from using too much soap. Certainly several people who suffered from spotty skins tell me they have had less spots since I advised them to go easy with soap.

The skin also contains sweat glands, and in accordance with Mr. Churchill's instructions, some of us are using them a bit more now than we used to. The article in question recommends you to shave under your arms, and seal the sweat glands up with a deo-dorant, while other steps are to be taken with other areas of the body. If you seal up the pores, you won't stop the glands from producing sweat. You will merely stop it getting out, which is not likely to do your skin any good.

As for our necks, we are told that soap and water isn't enough, you should mix powdered magnesia with rose water to make a paste, and leave it on for ten minutes. Your underclothes should be changed every day, your feet treated with talc powder and permanganate. Apart from soap and hair wash, which are certainly needed, though not in vast quantities, I counted nineteen articles recommended in a single page which may or may not improve a woman's appearance (I think rouge and eyebrow plucking very rarely do so) but are either of no use, or worse than useless, for health. I think some of the anti-squanderbug advertisements go rather too far, but when I read this article I felt as if I had fallen into a nest of squanderbugs. The magazine containing it was of course full of advertisements of cosmetics and aids to health and beauty. If an article like the one I am writing had appeared in it the advertisers would have objected. In 1938, of course, they objected to any articles crabbing Mr. Chamberlain's visits to Berchtesgaden and Munich, because the mere thought that war was possible would have wrecked the Christmas trade. Today they object to anti-waste propaganda which hits their favourite form of waste.

Very likely the woman who wrote the article believed every word of it. That is my main objection to advertisements. People believe lies, however outrageous, if you tell them often enough, as Hitler points out in *Mein Kampf*. As a physiologist I object particularly to lies about how our bodies work. A favourite bogy is uric acid, a non-poisonous substance which, however, accumulates in the joints in gout. As gout is a fairly rare disease, and there is no evidence that uric acid accumulates in ordinary rheumatism or arthritis, medicines which eliminate uric acid, or more accurately are alleged to do so, are of very slight value at best.

At the present time the cosmetic and patent medicine industry leads to a great waste of effort which could be used for winning the war and keeping up the people's health. If we can't have sugar on our cakes or cream with our pudding, it is ridiculous that we should be able to buy substances to plug our pores or flush our kidneys, which latter is easily done with a drink of water.

Unfortunately the interests concerned in selling such things are well dug in, and have influential connexions in both Houses of Parliament. Mr. Amery, for example, is, or was till recently, a

large shareholder in Beecham's Trust.[1] Any attempt to clamp
them down, even at the crisis of the war, would lead to howls
about bureaucratic and medical tyranny. But at least we can show
up their propaganda when it pretends to be propaganda for
health and efficiency, and can point out that a good deal of the
grease used in cosmetics is obtained from the traps of drains.

A socialist society would supply cosmetics to those who want
them, as the Soviet Union did in peace-time. But it would not
tolerate attempts to frighten people into using them even in
peace-time, and far less so during a war when waste is criminal.

Measuring Human Needs

The simplest formulation of socialism is "From each according
to his ability, to each according to his work," that of communism. "From each according to his ability, to each according
to his needs." Some people think it would be impossible to
determine needs, and that therefore communism is impracticable.

For example, Bernard Shaw, who on occasion describes himself as a communist, would like to cut the knot by giving every
man, woman, and child exactly the same income.[2] However,
Joseph Stalin, who has a somewhat better title than Shaw to be a
communist, said in his report to the seventeenth Congress of the
communist party of the Soviet Union, "Marxism starts out with
the assumption that people's tastes and requirements are not and
cannot be equal in quality or quantity, either in the period of
socialism or in the period of communism."

Any biologist must at least find the communist slogan interesting, because biology is becoming more and more a matter of
ascertaining needs. Some workers are concerned with nutritional
needs. They feed rats on simplified diets and find that their
growth slows down or they develop some symptoms such as
sore skins or sterility. Then something is added to the diet, and
the rats resume normal growth or the symptoms clear up. The
rats' needs are a fairly good guide to human ones, though not a

---

[1] See *Tory M.P.*, by S. Maxey, 1938.
[2] Since this was written he has changed his view.

complete one. For example, men, but not rats, get scurvy if their diet lacks ascorbic acid (vitamin C). Rats lose weight if deprived of a protein constituent called histidine, and men do not. Nevertheless, our present very successful rationing system is largely based on experiments on rats.

Other biologists study the needs of animals harmful to man. For example, the mosquitoes which carry malaria and yellow fever pass the larval stage of their lives in water. To control them we must know what sort of water each species needs. Can it live in shallow water, in running water, in brackish water, in water shaded from sunlight, and so forth? Some insects which destroy grain and other stored foods need a fair amount of humidity, others can live on dry food. But all need a temperature somewhere between freezing point and about 110° F., though many can stand high temperatures for a few hours, if the air is dry. One method of pest control is to see that pests do not get their needs.

Human physiological needs differ. A coalminer needs more food than a clerk, but he does not need exercise several times a week after work to keep fit. A sedentary factory worker has least fatigue and fewest accidents at a temperature between 60° and 65° F. A man on hard work is better off at about 55° F.

We have a scientific basis for assessing needs as to diet and many conditions in factories, but much less as to conditions in the home. Let us take a simple example. No doubt living rooms should get some sun. But is there any harm in having bedroom windows facing north? Or does sunshine in the day make them healthier at night? Nobody knows. But we do know that overcrowding to a density of more than one per room encourages a number of diseases. We know very little about needs for physical recreation. They may differ greatly in individuals, and I am sure that the compulsory "games" at many schools are overdone. But you must be careful in accepting statements from men who say they never took systematic exercise. My father used to say he never played games at the university, but on cross-examination admitted that he walked home and back—45 miles each way—at most week-ends. The plain fact is that we do not know our needs in this, and many other important matters. Only scientific investigation on a very big scale will determine them. This is one of the very many reasons why communism is impossible without science.

When we come to needs beyond those which a biologist can assess, the difficulties are obviously greater. But I believe I could persuade most readers that my needs, if I am to be as efficient as possible, are rather greater than those of a coalminer. I do not need so much food or so many clothes, or a daily bath, though I give myself this modest luxury in peace-time. But I do need a room with a large number of books, where I can work in the evening. I have to travel to keep in touch with scientific colleagues both in this country and others, quite apart from speech-making. If I am to visit my fellow academicians in Moscow in winter, as I hope to, I shall even want a fur-lined coat. Doubtless under communism I should not need many things which I have had to buy in the past, for example, a stiff shirt if I am to give an evening lecture at the Royal Institution and a top hat for funerals. But I should legitimately get a rather larger share of the national income than the average, whether in money or its equivalent, or in the shape of a free flat, books, motor car, railway tickets, and so on.

It is worth noting, however, that a sick man, woman, or child, suffering from a disease whose treatment is difficult, may need more man-hours per week than a professor and would get them under communism, just as in a well-run family a sick member will get more than an equal share of the family budget.

Clearly there will be difficulties in assessing needs under communism, and miners and professors may each think the others get too much, but many of these difficulties will be solved by the advance of science, and most of the rest on a basis of common sense.

Science and technology have made an age of plenty for all quite possible. This can be achieved under socialism. The Soviet Union was so backward technically at the time of the revolution that it was only entering the age of universal plenty in 1941. Britain could enter it within a year of establishing socialism. The transition to communism will probably start in the Soviet Union in a few years from the end of the war, with the free distribution of some necessities such as bread. It could not start so soon in

Britain even if we became a socialist country in the next ten years. For the younger people in the Soviet Union, brought up under socialism, mostly take it for granted that it is pleasant and honourable to work for the community. They are therefore ripe for communism. We shall not be so till socialism has taught us the same moral lesson.

Beyond the Microscope

The various sciences depend very largely on technical improvements. Mathematics could not go very far when XLIX multiplied by XCIV made MMMMDCVI. The invention (an Indian one, by the way) of our present numeral system made it possible to write this as $94 \times 49 = 4606$. The telescope made modern astronomy possible. Physics relies on hundreds of machines for accurate measurement, including the balance, the clock, the ammeter, the thermometer, the photoelectric cell, and others, many of which are familiar in industry. Although biologists may use any of the methods of physics and chemistry, the biologists' most important instrument is the microscope.

This has not only revealed hundreds of thousands of animal and plant species too small to be seen without it. It has shown that the larger ones are built up of cells whose structure is not very different in a man and a beetle, or in an oak and a moss. The cell is not merely a unit of structure, but of function. Isolated cells will live for a long time, and sometimes reproduce while smaller parts will not.

It is lucky for biologists that few cells are smaller than a wavelength of light, and most much bigger. For the microscope will not disclose details finer than this length. Quite recently we have got much greater precision with beams of electrons, but the new technique is still more difficult than microscopy two hundred years ago, and has not yet taught us much. So we have to rely on indirect evidence for finer details of structure. Probably my greatest contribution to science has been the interpretation of some of this evidence. At first sight it would seem that a hunt in north-east London for the relatives of a man who is both colour-blind and haemophilic (that is to say whose blood will not clot) could hardly tell us about structures inside the cell too small to be seen with the microscope. But it did.

Every cell in a human being, an animal, or a plant, contains a nucleus. When the cell divides the nuclear material is organized into threads called chromosomes, each of which divides in two, half going to each daughter cell. The cell divisions connected with reproduction are rather different from the normal. But the net result of them, and of the fusion of a female and male cell in fertilization, is that an individual usually gets half his or her chromosomes from the mother, and half from the father. The chromosomes consist, in part, of smaller structures called genes, which are the material basis of heredity. For example, a white cat is white (and often deaf) because it has a gene which is not there in ordinary cats. Unless both its parents were white, it only has one such gene, and when mated to an ordinary black or tabby, it hands it on to half its kittens on an average.

The American biologist Bridges, who, by the way, was a keen co-operator, and in other ways too radical for many of his contemporaries, first proved, from a study of flies in which the number of chromosomes were abnormal, and well-known genes were inherited in an abnormal way, that genes were carried by the chromosomes. Only rarely can a gene be seen even with a powerful microscope. But a few are visible as alterations in the pattern of a chromosome, in insects where the chromosomes in certain cells become very large. None have yet been seen in vertebrates, let alone man, but we are beginning to know where they are situated, by the study of what is called linkage.

Genes in different chromosomes are inherited independently. For example, a man who has inherited the genes for the Rh substance in the blood[1] and for hay fever or other allergic diseases from his father, hands them down independently to his children. Those who inherit the Rh antigen are no more likely than the others to develop hay fever.

But where the genes are on the same chromosome this is not so. If a man is haemophilic and colour-blind his children are normal, but half his daughter's sons are colour-blind, and half haemophilic. And it is usually the colour-blind ones who are

[1] See p. 24.

haemophilic. On the other hand, if a woman inherits haemophilia from her mother and colour-blindness from her father, the sons will almost all be either haemophilic or colour-blind. Only a very few will have both defects or neither. The nearer two genes are in a chromosome, the less likely they are to separate if they came in from the same parents, or to go to the same child if they came from separate parents.

On this principle maps of the chromosomes have been made in a few animals and plants. These maps give us a detailed picture of organization inside the cell, on a scale too small to be seen with a microscope. On one small section of one human chromosome I have located five genes, and other workers two more. It is going to take centuries to map all the human chromosomes. When we have the maps we shall be able to say something like this. Here is a gene which in normal people produces a substance concerned in blood clotting, and which is out of order in haemophilics. One hundred-thousandth of a millimetre away is the gene whose failure causes colour-blindness, and in between them the gene whose failure causes a particular form of paralysis. When we can do this, eugenics will become scientific, rather than class propaganda, as it generally is at present. But we shall also have gained what may be very much more important, a knowledge of the internal organization of our cells as detailed as the knowledge of anatomy which guides both the surgeon and the first-aid worker. We may find out how to get this knowledge by less roundabout methods. I hope we shall.

Now the war is over, I am beginning to start work again on these problems, though I cannot do much till University College is rebuilt. The Nazis have done such horrible things on the basis of their false theories of heredity that all work on this subject is inevitably suspect. But we can best counter these falsehoods by discovering the truth, and the truth is going to take the study of human, animal, and plant structure a whole stage beyond that which the microscope made possible.

Evolution, and Our Weak Points

My wife has just started a raging cold. By the time this article is printed I shall probably have one too. Worse, we shall probably be spreading them to others if we go into public shelters. Like many others, I get one or two of these acute nose infections every winter, whereas all my other organs together do not go seriously wrong once a year. Why is the nose such a weak point?

Part of the answer is to be found in the history of human evolution. Compare a man's nose with that of any other mammal, say a horse, dog, or rabbit. The air going into and out of its lungs has a nearly straight run from the lungs in and out through the nose. If there is a bend it is where the head joins the neck. In man each nostril takes a hairpin bend, with very awkward results. When a dog sneezes, the air gets a straight run, and he can clear his nose. A human sneeze cannot get through the narrow and twisted nostrils, so when we sneeze we have to open our mouths. This twisting of the nostrils is a result of the great growth of the brain, which has distorted our heads from the usual mammalian shape. More accurately, we have kept a type of head shape which is common in embryo mammals, but which is generally liquidated long before birth.

Our mouth is another weak point. Our teeth suffer from gross overcrowding, and that is one reason why they so often decay. We are probably evolving towards a condition with fewer teeth, for a good many people never cut their wisdom teeth, and perhaps none of our descendants will do so. Other organs in the face suffer from this distortion. There are a number of cavities in the skull filled with air and communicating with the nose. These may become inflamed, or their communications with the nose blocked. The resulting sinus or antrum disease is almost always painful and sometimes fatal.

Still another set of human ills come from the fact that only in the last million years or so have our ancestors taken to standing on their hind legs. Our feet go wrong more often than our hands. That is why chiropody is sometimes a necessity, while manicure is a luxury.

But the effects of the erect posture are more serious elsewhere. A dog's internal organs are suspended from above, that is to say from his back. This form of suspension has only been partly modified in human beings. So they are very liable to slip in a direction which is downwards in a man, but would be backwards in a dog. This tendency results in a variety of complaints including rupture and prolapse of the womb, and accounts for some of our digestive troubles. The narrowing of the orifice between the bones of the pelvis, whilst it checks this tendency, exposes women to more pain in childbirth than other female mammals. The heart is much higher above the ground in man than in a four-footed animal of the same size, and in consequence, in order that the blood should be able to return to it, the pressure in the leg veins must be much higher. When they cannot stand up to this high pressure we get varicose veins.

In fact, as a result of standing up, our bodies have developed internal contradictions, just as our society has developed them as a result of improvements in productive forces. Unfortunately we do not yet know how to change our bodily build as some of us know how to change the social order. So our descendants will probably suffer from hammer toes and varicose veins after they have ceased to suffer from capitalism.

In just the same way our brains are so organized as to dispose us from time to time to conduct and emotions more suited to animals than to human beings. But here the situation is very different to that of our anatomy. You can make a plant grow into many different shapes, but you can't alter the human form greatly, except by mutilation.

Some animals, particularly insects, have very stereotyped forms of behaviour. We may say, if we like, that they have fixed instincts. Mammals and birds, however, are much more plastic in their conduct, and some can learn a good deal. As regards conduct, human beings are far more plastic than any animals. This is one of the things which makes them human.

Some eugenists hope that by selective breeding it may be possible to produce a human race in emotional harmony with its environment. Trotsky adopted this view, which is certainly un-Marxist. For a society composed of such people would not

progress. Animals which are very well adapted to some particular environment seem to be dead ends in evolution. They may
survive for millions of years without much alteration, only to be
swept away when a large change in climate or vegetation occurs.
Their place is then taken by other animals which have been less
specialized, and temporarily less successful.

In the same way if men and women were ever perfectly adapted
to a society, then that society could never undergo fruitful
change, though it would perhaps decay very slowly. Technical
improvements which might upset the social equilibrium would be
resisted, as the mediaeval guildsmen tried to prevent new processes of manufacture. Engels pointed out that "it is precisely the
wicked passions of man, greed and lust for power, which, since
the emergence of class antagonisms, serve as levers for historical
development." So until society has been perfected it is futile to
wish that individuals should be so, even if this were possible.

No doubt when the classless society has been achieved our
descendants will have to tackle their physical and moral imperfections. But I think they will do so much more by changing the
environment than by the very slow process of eugenical selection, even if the latter is used to some extent. Man has managed
to live in cold countries by inventing clothing, fire, and houses,
not by growing thick hair like polar bears. Children are made
into social beings by education, not by breeding from those who
behave best. And as we learn about the physiology of development we may well be able to correct the anatomical defects of
which I have spoken without surgical operations on the one
hand, or eugenics on the other.

Man's Ancestors

When Darwin wrote *The Descent of Man* he was mainly concerned to show that men were descended from animals fairly like
modern apes, that these were descended from reptiles, reptiles
from fish, and so on. Today we have much more evidence in
favour of this theory, and also vastly more details of the pedigree.

First of all palaeontologists have now unearthed perhaps fifty times as many fossil vertebrate species as were known when Darwin wrote, and have studied them much more closely, for example, by examining bone and teeth sections with the microscope. Secondly, we know more about the relationships of living animals, because we have studied their development in greater detail. Thus it is obvious that our finger bones correspond to those of a lizard, but only embryology showed that the small bones which conduct sound from our ear drums to our internal ears correspond to relatively large bones in a lizard's jaws. We can also compare animals on the basis of their chemistry as well as their anatomy. And we can compare microscopical structures in the cells.

Finally we know a good deal more than Darwin did about how evolution took place. Geneticists have now produced varieties of a number of animals and plants which give sterile hybrids when crossed with the original type, as the horse does with the donkey. The fact that all varieties of dog or rabbit can be crossed was used as an argument against Darwin to show that species are quite different from varieties.

The earliest ancestors of which we know anything were something between worms and fish. They lived about 500 million years ago, and had no jaws, but round mouths, and a number of bony or gristly rods supporting gills round their throats. They were probably descended from worm-like animals, of which traces have been found in earlier rocks; but we have not enough fossils to be sure.

The first great step in evolution was the hingeing of one gill arch to form a lower jaw, so that the animal could snap instead of guzzling mud or sucking motionless food on the sea bottom. With this went a development of paired fins for swimming, and doubtless improvements in the sense organs and brain. By and by a second gill arch was added to strengthen the lower jaw. This step was only discovered by Watson in London just before the war. The paired fins developed a narrow base, and became more like flippers. A pocket off the gullet could be used for holding air, either for buoyancy or to help breathing if the water was foul. Our ancestors were definitely fish, but less stream-lined than many modern forms.

Then in the Devonian period, when there were many shallow lagoons, our ancestors came out of water. Fish still do it. In the Zoo aquarium before the war one could see the mud-skipper, a tropical fish which can go some distance out of the water. It has developed eyes for seeing in the air, and joints in its fins. Unfortunately no one told it that it was 250 million years late in its attempt to colonize the land. The next stages included the development of legs and of nostrils on top of the head, and our ancestors at the time when the coal was formed were still amphibians like newts, almost certainly with slimy skins and probably passing their infancy in the water.

After this they became complete land animals not very unlike lizards, with dry skins, and laying eggs on land, so that for the first time they were almost independent of water. One group of reptiles, the theriomorphs, seem to have had hair instead of scales from an early date, because we find holes on their face bones similar to those in which the whiskers of many mammals are set. They began to lift their bellies off the ground, and to resemble mammals in various ways. No one yet knows when and where they took the great steps of becoming warm-blooded, and bringing forth their young alive. We shall know about the latter point when more fossils of pregnant females are found, as fossils of the pregnant ichthyosaurus, a whale-like reptile, have been. Nor do we know when they began suckling their young.

Animals which suckle their young have at least the beginnings of social life. During the last 30 million years their most striking advance has been in brain size, though they have produced specialized forms, runners like the horse, gnawers like the rat, burrowers like the mole, swimmers like the whale and seal, even the flying bats. But man is rather a primitive mammal anatomically. We have lost less than most, tails, tactile whiskers, a good deal of hair, and a few teeth. We have kept all our fingers and toes, and our collar-bones. Our main specializations are in our large brains, our highly developed hands, our close-set eyes, and our heels.

Our immediate ancestors were climbers fairly like some existing monkeys, but our structure has not changed much in the last half million years. Since our ancestors discovered fire and began to co-operate in production, our main evolution has been social, and there is no need to tell readers of this book that human society is still imperfect, and changing very rapidly.

There were *Giants*

We do not know where and when men originated. In one sense we are never likely to know, for the process was probably not quite a sudden one. If, with Engels, we make production for future use the test of manhood, it would be hard to draw the line between an animal which occasionally sharpened a stick or chipped a stone, and a man who did so habitually, even if we knew all the facts.

We shall certainly not be in a position to give a definite answer until most of Europe, Asia, and Africa have been searched for human remains at least as carefully as Britain and France have been searched at present. But on the present evidence the most likely site seems to be South-Eastern Asia or the Malay Islands, and the time before the ice ages. *Pithecanthropus*, a very primitive human type found in Java fifty years ago, and *Sinanthropus* more recently found near Peking, both come from this area, and the latter was certainly a man on Engels' definition, as he used fire. However, he had a good many ape-like features.

In May of this year Dr. Weidenreich informed the American Ethnological Society of discoveries made in Java in 1939 to 1941 by Dr. von Koenigswald, of the Netherlands East Indies Geological Service, who has been missing since the Japanese conquest of Java. A preliminary account has been published in *Science*.

In the volcanic ash beds of Trinil in central Java he, or rather his Javanese collectors, found a series of skulls and lower jaws, which are definitely human, though primitive, and some of which are enormously larger than those of any living or previously described fossil men. The most complete skull, for example, had room for a bigger brain than any ape's, though a small one by modern human standards. But the head must have been of human size because of the thickness of the bone, and the presence of a ridge on the top of the skull to which great jaw muscles were probably attached as in the gorilla. The upper jaw was so big that it allowed of a gap between the canine and incisor teeth, but the canines were of the human type, not like the tusks of many apes. This form has been called *Pithecanthropus robustus.*

A fragment of a lower jaw belonged to a much larger type, which was probably about the size of a large male gorilla, but yet in shape far more like a man's than a gorilla's. Its owner has been called *Meganthropus palaeojavanicus.*

Finally in Chinese apothecaries' shops in Hongkong, von Koenigswald bought three molar teeth of human though primitive pattern, and still larger size. The volume of the crowns is about six times the volume of the crown of the corresponding tooth of modern man, and twice that of a male gorilla. If the rest of the body was in the same proportion, its owner may have weighed half a ton or so. These teeth probably came from caves in Szechuan, Yunnan, or Kwangsi. If so, excavation on scientific lines should reveal complete skeletons of these giants, or at least thigh-bones like that of *Pithecanthropus*, which made it sure that he walked upright, and gave a rough idea of his height.

I do not think that these remains can be used to explain the legends of giants found in so many ancient books, such as the Bible, the Mabinogion, and the Edda. These were composed two or three thousand years ago at most, while the human or near-human giant fossils probably date back half a million years.

The most striking fact about these fossils is that the largest are the most primitive, that is to say show the most ape-like features. Dr. Weidenreich thinks that they were quite probably in the direct line of human evolution. If so, one of the teeth which von Koenigswald bought in Hongkong may have belonged to my great .... grandfather (with about 20,000 "greats" in the gap). If he or she was an ancestor of any living man he was also, of course, the ancestor of all living men. Most palaeontologists are inclined to think that the giants were a side line, cousins rather than ancestors.

If these giants were our progenitors there was a long period during which our ancestors were getting smaller. This makes a great deal about human evolution a lot easier to understand. It was not clear why men had lost most of their hair, and their canine tusks, before they began to make weapons or use fire. If they were giants it is quite intelligible. A giant animal in a warm country has difficulty in keeping cool, and generally loses most of its hair, like the elephant, the hippopotamus, and the rhinoceros. An animal which could tear up a tiger with its hands would not need dog-teeth. As men got smaller they would need weapons, fire, and probably greater sociability to allow them to combine for hunting and defence against wild animals. All this is, of course, speculative. The great development of biology in China under such men as Professor (now General) Lim makes it fairly sure that many of the gaps in our knowledge will be filled in the next dozen years.

The whole episode is typical of the progress of science. An utterly unexpected fact has turned up, as unexpected as the activity of radium, or the difference between the chromosomes of the sexes. It will mean a certain alteration in current theories, though rather a shift of emphasis than a thorough revision. In the long run it will be fitted in, and will probably make the previously known facts easier to understand.

4

MEDICINE

The Common Cold

THIS is the worst season of the year for common colds. From one to three or more times a winter, most of us are inefficient for several days as a result of this infection. In normal times the most public-spirited thing to do if you had a cold was to knock off work, so as not to affect your mates. At present I think we are supposed to stay at work and trap the germs in our handkerchiefs. The *Lancet* recommends us to wear a pad over our noses and mouths for three days, but it is not so easy to get the material.

There is no doubt that a "cold" is due to infection. This has been shown in a great many ways. Careful studies have been made in small islands where no one has had a cold for many months until a ship arrived, and then it has gone round the whole population. The agent is too small to be seen even with a microscope, for colds have been given, both to men and apes, by filtrates of nasal secretions which had passed through a filter so fine as to stop all bacteria. Apes get our colds, and get them very badly. The glass screens between us and them in the Zoo are to protect them from our airborne diseases, including tuberculosis, but particularly colds.

While a virus is one cause of colds, it is not the only cause. The weather is somehow concerned, which is why colds almost disappear in summer. But it is not the main cause. Arctic explorers never get colds, until they come back to "civilization" and get real bad ones.

Nobody knows where the virus of the common cold spends the summer. Even in warm weather there may be enough people with colds to keep it going. Perhaps a few people can carry it without showing any symptoms. Until this is known there is no prospect of wiping out colds.

Attempts to prevent them with vaccines, serum, and so on, have been a failure, or at least have not been very successful. An English professor injected half his medical students with a vaccine which was supposed to be prophylactic. He asked them if it had done them good, and the majority said yes. But he also made them keep diaries of their colds, and found that they got just as many, and as bad ones, as the untreated half. Other workers have of course claimed better results.

Colds can probably be cured with some of the new drugs related to sulphanilamide. But these are dangerous substances, and can cause illness far worse than a cold. The risk is worth taking in the case of blood poisoning, pneumonia, or gonorrhoea; it is emphatically not worth taking for a cold.

However, a very great deal can be done for a cold with ephedrine. This drug is derived from a root which has been used in China for a long time under the name of Ma Huang, *Ephedra* being the scientific name for the plant. Many traditional Chinese medicines are quite worthless, as are most of the traditional European ones, including a good many which are still prescribed by doctors and sold by manufacturers. Ma Huang is rather uncertain in its action, as the amount of ephedrine in different roots varies. This is generally so with herbal remedies.

Chinese and Japanese scientists have investigated a number of these medicines, and the most valuable substance so far found in them is ephedrine. Its chemical formula is similar to that of adrenaline, the substance which the adrenal glands, lying close to our kidneys, pour into the blood during violent emotion or exercise. Adrenaline, among other things, contracts the small arteries and speeds up the heart. But its effects do not last very long, or our hearts would go on thumping for hours after we had keyed ourselves up when we run to catch a train or hear the sirens announce an alert. We have chemical means of destroying it in a few minutes, and making more when it is needed.

Ephedrine has most of the effects of adrenaline, but we do not destroy it rapidly, so they last for some hours. It can be used locally, as surgeons use adrenaline to stop bleeding from small blood vessels. A number of solutions of ephedrine for dropping into the nose are on the market, and some of them are pretty effective. If you treat your nose with one of them before going to bed the blood vessels in the inflamed membrane shut up, and I, for one, can breathe through my nose again. This not only enables me to go to sleep. It helps me to sleep with my mouth shut so that I do not breathe cold air, and the infection is much less likely to spread to my throat and lungs. It also stops me snoring.

One can also eat ephedrine. I do so when I have a cold, and it certainly stops my nose running. But it puts my pulse rate up to 100 or so per minute, and raises my blood pressure. As my blood pressure is normally very low, I don't mind; but no one should eat it unless they are sure that their blood pressure is normal or low, and I don't recommend anyone to start on more than half a grain. Ephedrine is also liable to keep you awake if you take it by the mouth. But I certainly recommend it. I made two speeches on January 30th and 31st, and I doubt if any of my audiences spotted that I had a fairly bad cold—or should have had but for ephedrine.

Very likely some other compound will be made which, for the same action on the nose, has less effect on the heart than ephedrine, and less on the brain than benzedrine, which is also of some value against colds. Obviously, systematic work on these lines will not be done during the war.

And even in peace the medical profession is not much concerned with colds. They are not consulted about them, and as we only pay them when we get ill, they can hardly be expected to do much research on the subject. Yet colds cause a vast amount of unhappiness and inefficiency. When the economic basis of the medical profession is changed so that their main function becomes, not merely to cure or prevent serious disease, but to keep us in perfect health, colds and many minor illnesses will be attacked as vigorously as typhoid or smallpox. But this is hardly likely to happen until medicine is socialized.

Moulds versus Bacteria

The press has recently been full of stories about a new and wonderful cold cure, and one newspaper at least has been urging its mass production. The substance in question is called patulin and is made by a mould called *Penicillium patulum*. It is a fairly simple organic compound with only seven carbon atoms, and can be purified by crystallization.

Another product of a different mould species, penicillin, has proved extremely useful in stopping a variety of acute infections, and is being used in a big way for the treatment of septic wounds. Its efficiency rivals that of sulphonamide and its derivatives.

These substances are not antiseptics in the ordinary sense. When added to a culture of bacteria they do not kill them, as mercuric chloride or phenol do. They stop them growing, and are called bacteriostatics. Many of them have similar effects on other living cells. If we injected mercuric chloride into a man with septicaemia we might kill all the bacteria in him. But if so we should kill him too. A bacteriostatic stops the bacteria from dividing. It may also stop human cells from dividing for a while, but if so no great harm is usually done, though there is often a danger of anaemia, since the cells in human blood have to be replaced fairly quickly, and any check on cell division may lead to a shortage of blood cells.

We know how some of the bacteriostatics act. Before any foodstuff can be used by a living organism it must unite with a special molecule of protein called an enzyme which causes a chemical change in it, the first of a series of changes in which it is either built up into living substance or used as fuel for muscular work or heat production. For example, sugar is combined with phosphoric acid by a special enzyme before it is used for muscular work. An enzyme will not only unite with a foodstuff, but with other substances of similar composition. If it cannot change them, its action is more or less completely blocked, and the cell containing it is starved of a certain product.

Much the same thing happens when human beings are given a bogus food or drink, for example "lemonade" made from citric acid, not lemons, or a fat which has been heated so as to destroy the vitamins which it contains. Our appetite is satisfied, but we do not get the benefit which we should normally get. At present most of our foodstuffs are pretty good from this point of view. But "decontrol" and "free competition" after the war, if they are applied to food, will mean a shortage of essential foodstuffs once again.

To go back to bacteriostatics, the sulphonamide derivatives act by jamming an enzyme which uses *p*-aminobenzoic acid, one of the substances which used to be lumped together as "vitamin B." It acts by producing the equivalent of a vitamin deficiency, which is much more serious for bacteria, which may divide several times an hour, than for human cells, most of which do not divide at all.

Penicillin was discovered from the observation by Fleming that some moulds prevented the growth of bacteria in their neighbourhood. This may be the function of such substances in the mould's life, since bacteria compete with moulds for rotting foodstuffs.

Dr. E. W. Gye, of the Imperial Cancer Research Fund's Laboratory, wished to try the effect upon cancer of some of the simpler substances made by moulds. It is obvious that they might slow down the abnormally rapid growth of a cancer, without stopping all growth in normal parts of the body. He got patulin from its discoverer, Professor Raistrick, of the London School of Hygiene and Tropical Medicine. Having a bad cold, he tried washing his nose with it. The solution was rather too strong, and hurt him a bit, but his cold vanished. Some colleagues had equally good results. It was next tried by Surgeon-Commander W. A. Hopkins, at a naval depot. He found that fifty-five out of ninety-five men whose noses were washed with it had recovered in twenty-four hours (besides a number who said they were cured but were not). Only eight out of eighty-five untreated men recovered within the same time.

So patulin certainly won't cure all colds. Moreover, some of those whose noses had ceased running developed sinusitis, that is to say, pains in the face, due to inflammation of hollows in the bone connected with the nose. Even more important, two other doctors who tried it got no good effect at all. And no one claims that it works against influenza.

This does not prove that patulin is a fraud or a delusion. The difficulty may be that we have lumped together a number of different diseases under the name of "common colds," and are then surprised that what stops one does nothing to another. When we know more about colds this may seem as silly as to expect that the same treatment is best for potatoes and beans, because they are both vegetables. Very likely the colds at the naval depot went round the men very quickly, and half of them were due to a single germ or group of germs which is affected by patulin, while most of the other colds were due to a different agent. At present patulin is being tested by a number of doctors, and by next spring we should know whether it is worth while distributing it widely. My own guess is that it will cure some colds, and that other substances will cure others, so that it will ultimately be possible to make up a mixture that will cure most.

Till then I shall go on treating my own colds with ephedrine, which prevents my nose running, but unfortunately has other less pleasant effects, and would certainly shorten some peoples' lives if they took enough to deal with their colds. So I don't recommend it for general use.

Venereal Diseases

The number of people affected with venereal diseases is increasing, as it always does during war. And there is a great diversity of opinion regarding the best means of fighting them. This arises partly from prudery, but largely from sheer ignorance.

What are venereal diseases? This name is given to a group of diseases which are usually (though not always) passed from one person to another by sexual intercourse. It is most important to realize that they can be passed on in other ways. An infected mother commonly gives them to her children, and a towel or washbasin may carry the infection.

Some religious people regard them as a punishment for sin. This is certainly not the Christian doctrine, as can be seen from the ninth verse of chapter three of St. John's Epistle. It is also untrue. A wedding-ring does not prevent venereal infection. It would be truer to say that they were punishments of ignorance.

A number of diseases are passed from one animal to another in this way. One is known in horses, another in dogs. In the human species there are two main diseases. The worst is called syphilis, or great pox, and is due to a microscopic corkscrew-shaped germ.

This generally enters through a scratch or raw place in the skin. After about four weeks a hard but painless sore develops. If this is treated at once the patient need never be ill at all. If not, he or she gets swollen glands and a variety of eruptions on the skin, and others in internal organs which may cause all kinds of symptoms from sore throat to madness.

These generally die down, but after several years a new set of symptoms develop. They include "bad legs," erosions of the bones, and weak spots in arteries which may burst and cause sudden death. Finally, sometimes after twenty years, the nervous system may be affected. The first symptom is often a loss of feeling in the feet. The patients lift them high up and stamp when walking, and are sometimes thought to be paralysed. They often end up as the rather cheerful kind of lunatics who sign cheques for a million pounds, and say they are Jean Harlow or Winston Churchill. An infected mother commonly infects her babies. Some die before birth. Others are born with the disease, and this congenital syphilis is one of the main causes of mental deficiency.

Gonorrhoea, or clap, is a more local infection due to a coccus not very unlike those causing boils and blood poisoning. It starts within a week with a discharge of pus from one or other of the tubes opening at the place of infection. At this stage it can be rapidly and completely cured. If not, it may spread inwards, causing great pain, and sterility in both sexes. It sometimes causes severe rheumatism and, if it reaches the eyes, blindness, especially in new-born babies of infected mothers. It is not such a great killer as syphilis, but is commoner.

Two other common diseases of this group, soft sore and lympho-granuloma, do not spread far from the point of infection, but the latter is very serious. Venereal diseases rank fairly high among the causes of death, but no one knows how many they kill, as deaths are commonly registered as diseases of the liver, blood vessels, or whatever organ first breaks down.

These diseases could and should be wiped off the face of the earth. They would be abolished if all patients were cured, as they could be, before infecting anyone else. But people are ashamed to ask for treatment, or do not believe that anything is seriously wrong until they are not only ill, but highly infectious. It will not be easy to educate people about these diseases. Fourteen years ago Moscow was full of posters showing their symptoms. They were disgusting—but not so disgusting as the reality. We prefer to hide the truth in this country.

The second line of attack is to cut down sexual promiscuity. Prostitution will never be abolished until the men concerned are punished as severely as the women, or as in the Soviet Union, more severely. Young people will be promiscuous as long as low wages, bad housing, means tests, and other economic causes make early marriage difficult. And older people will do the same until divorce is made a good deal easier when a marriage has failed. History shows that "pi-jaw" has very little effect. And no wonder, if moralists lump together as "sinners" a couple who live together without being married, and a man who deceives a girl and leaves her to look after the baby.

The third line is prophylaxis by antiseptics. There is very little danger of infection if the places where infection is likely are washed within an hour or so, first with soap and water, then with a bright pink solution of potassium permanganate, and then rubbed with calomel ointment. It is important to get the permanganate solution well into the tubular part of the organ concerned. This can be done with a syringe. Some other antiseptics can also be used.

When these substances are supplied and men or women instructed in their use, venereal infection is very rare. But they are not easy to get, and are generally sold without any instructions. Many people are violently opposed to prophylaxis, though it is not clear to me why it is wrong to wipe out the disease after two hours, but right to do so after two weeks.

At present a controversy is raging about Regulation 33B, under which, if two patients trace their infection to the same third person, he or she can be forced to undergo treatment. It has been said that this will lead to blackmail, and to further persecution of prostitutes by the police. This may be so, but as complaints can only come from infected people, I do not think either danger is very great.

We shall not abolish these diseases until everyone is educated about them, as they are about other infectious diseases, and until we have a social system which not merely discourages promiscuity, but encourages early marriage. Meanwhile, I believe that one of the best ways in which promiscuity and venereal disease could be kept down would be to allow more and longer leave to married members of the forces, both men and women.

Causes of Cancer

In the *Daily Worker* of October 28th Frank Lesser wrote that a Pensions appeal tribunal, dealing with the case of a soldier dying of cancer, said that "as the cause of cancer was not known, and as anybody might at any moment be found to suffer from it, there was no evidence that death was accelerated by conditions of service." The tribunal may have been right in their judgment in this particular case, but if they said what is reported, they were certainly wrong in their science. Cancer, like any other natural event, has a great many causes, and some of them are known.

Cancer is simply a disorderly growth of the cells of some part of the body, which invade other parts. Excessive growth in an orderly manner may be harmless, as with warts, or very dangerous, as in the brain, where the skull leaves no room for expansion. But in most parts of the body quite large tumours can be safely removed provided they do not spread. In the case of cancer, one may cut the original growth out or kill it with X-rays or radium, but unless this is done very soon, some cells from it may have been carried by the blood or lymph and settled down in some other parts of the body, or it may have spread in another way. So the best place to have cancer, if one must have it, is the skin or some easily accessible organ such as the breast. Most skin and breast cancers, if removed soon enough, do not spread.

If we are ever able to point to a single cause of cancer, it will almost certainly be by standing the question on its head, so to speak, and asking why every cell in the body does not occasionally divide. Some cells will go on dividing every day or two in a suitable fluid. We can only guess at what normally restrains them. Till we know this we can only say that a lot of different agencies favour the development of cancer, and may be regarded as causes.

The first, curiously enough, is hygiene. Every year of peace a larger proportion of all deaths is due to cancer, simply because people survive other dangers to die of it. Cancer is a disease of old age, and kills few people under forty-five, so if the deaths from phthisis or heart disease are reduced, more people live long enough to die of cancer. For this reason it is rather a commoner cause of death among the rich than the poor. Among 1,000 poor men aged sixty-five more will die of cancer within a year than among 1,000 rich. But so many more of the rich than of the poor live to be sixty-five, that this gives them a higher overall cancer death rate.

Poverty probably causes cancer by means of some form of dirt. For the extra deaths from cancer among poor men as compared to rich men of the same age are mostly from cancer of the skin, mouth, gullet, and stomach, where dirt can reach, and not of other internal organs where it cannot.

Motherhood lowers the chance that a woman will die of cancer of the breast, but increases her chance of dying of cancer of the womb. It is probably a help against breast cancer to suckle the baby, and it is also probable that the first two or three babies do not greatly heighten the risks of womb cancer, while large families do so. But statistics on these points are not conclusive.

Alcoholic drinks certainly increase the chance of cancer. The death rate from cancer among publicans, for example, is well above that of average men of the same age. There is good evidence that it is higher among habitual beer drinkers than the rest of the population, but it is not so sure whether wine or spirits have any such effect.

A group of chemical substances cause cancer, even in very small quantities. These are found in some types of lubricating oil, and were particularly common in the shale oil from which Lord Linlithgow's fortune was largely derived. They are also common in pitch, tar, and soot. So cancer, especially of the skin, was found to be common in workers with shale oil, tar, and pitch, briquette and patent fuel makers who use pitch, chimney sweeps, and certain workers exposed to mineral oil, particularly cotton mule spinners. The condition was found in London sweeps by Dr. Potts in the eighteenth century, and most of the research was done by doctors. But it was James Wignall, an organizer of the Dockers' Union, who discovered it in the South Wales patent fuel workers. The cancer-producing substances act slowly. It takes over ten years before a cancer develops in oil-soaked human skin. But they will act within a few months on mice. And by working on mice, Professor Kennaway, of the London cancer hospital, was able to isolate some of them, and to show that some lubricating oils are vastly more dangerous than others.

It is very possible that cancer spreads in man partly, at least, because the cancer makes a substance similar to those found in pitch and oils. For Shabad in Leningrad extracted an oily substance from the livers and lungs of men dying of cancer which caused it in animals, and his discovery was confirmed in Britain, America, and South Africa.

Other kinds of chronic irritation, for example, X-ray burns, and repeated burns with hot objects, can cause cancer. Skin cancer is common in Kashmir where the inhabitants keep warm by carrying pots of glowing charcoal under their clothes. Heredity also plays a part.

In fact we cannot speak of *the* cause of cancer. Yet we know enough of its causes to cut down the death rate considerably. Most of these causes take many years to act. So the tribunal was very likely correct in its verdict according to the existing rules. But just because the cause is often doubtful, one might hope that a tribunal would give a soldier's dependants the benefit of the doubt. If the tribunal is forbidden to do so, an ex-soldier like myself may be pardoned for thinking that its rules should be amended.

Until more is known about the causes of cancer, the best way of fighting it is to get a medical overhaul as soon as we feel any lump in any organ, or notice any abnormal bleeding, let alone chronic internal pains. Cancer is not always painful, especially in the early stages; and the fact that it can often be cut short is shown by the fact that doctors, who naturally die more than the average from infectious diseases, have a low death-rate from cancer.

A New Attack on Cancer

The most important drugs which have been brought into use in the last few years slow down the growth of bacteria without doing much harm to the human body. They fall into two classes. One consists of compounds related to sulphanilamide made in the laboratory. These act by blocking enzymes used by bacteria for their growth. They thus prevent the bacteria from growing in our bodies. They may also have similar effects on larger organisms. For example, if you give a dose of sulphanilamide to a hen it lays eggs with soft membranous shells for several days.

The other group consists of compounds such as penicillin made by moulds. The composition of some of these is known, but that of penicillin is not—or if it is, it is an official secret. These also stop bacterial growth, but we do not know how. In any case, an explanation of how they act could only be given in terms of organic chemistry, and would fill many pages. But another group of substances is coming into medical use, whose action can be explained more easily.

These are the artificial radio-active elements. Fifty years ago almost everyone but dialectical materialists thought either that the known chemical elements were created by God, or had existed eternally. Even idealists who did not believe in matter said they were "necessary forms of thought" or something equally ridiculous.

We now know that there are many hundreds of kinds of atoms, but that most kinds are unstable. Naturally enough, the stabler kinds were discovered first. Perhaps no kinds of atom are quite stable, but many kinds seem to last for many thousands of millions of years on an average, while the most unstable ones may have an average life of a fraction of a second. The unstable atoms were only discovered because they shoot out rays or particles during their transformation. At first the only ones known were heavy ones such as radium and uranium, which occur naturally. It was soon found that they had a value in the treatment of cancer.

For example, cancer of the breast is quite definitely curable if tackled soon enough. In the early stages it is often enough to remove the swelling. Later on there is a chance that cells from it may have spread, and needles containing radio-active elements are pushed into the organs where it is likely that cancer cells will be found. The particles and rays from radio-active substances have little effect on cells except when they are dividing. The cancer cells divide much more frequently than ordinary cells, which is why cancer is deadly. But in consequence they are more easily killed by radium and X-rays. So if the needles are in the right places and left there for some days, there is a good chance of killing all the cancer cells without killing too many others.

Recently it has been possible to make radio-active forms of the lighter elements which are part of the normal make-up of the body by speeding up the nuclei of the heavy kind of hydrogen in a very powerful electric field, and shooting them into another element. This is most efficiently done with a machine called the cyclotron, invented by E. O. Lawrence, a Californian. For example, by firing at a target covered with phosphorus, about one phosphorus atom in a million can be made radio-active.

These artificial radio-active elements behave chemically like the corresponding stable ones. In particular, radio-active phosphorus is taken up by bones and by rapidly growing cancers, and in consequence, if the patient is given a drink of radio-active phosphate, it is automatically concentrated where it is most likely to be of use, and it seems to have cured some cancers. Dr. Lawrence, the brother of the inventor of the cyclotron, has used it with marked success in the usually fatal disease of leukaemia, in which the bone marrow or lymph glands produce too many white corpuscles. He has been particularly successful where the bone marrow was responsible, as the cells which were dividing too often were bombarded with rays from the bones surrounding them.

Similarly, radio-active iodine has been very successful in slowing down over-production of its hormone by the thyroid gland, though it has not been successful, so far as I know, with cancers of the organ. A number of other elements are being tried out, and probably a great deal has been learned in the last year. The ideal method would be to find some special compound which is taken up by cancer cells from the blood and make it up with a radio-active element. The trouble here is that all cancers do not behave alike, in fact it would be nearer the truth to say that no two cancers behave in the same way.

Not only has the use of radio-active elements given us a hopeful method of dealing with cancer, but it has made all kinds of physiological experiments possible which were formerly quite out of the question. For example, unless one takes a very large dose of common salt, the amount in the blood stays so nearly steady that no rise can be detected. But if a man drinks a solution of common salt containing radio-active sodium, and puts his finger on an instrument called a Geiger counter, this will begin to register radio-activity in a very few minutes. There is enough radio-active sodium in the blood to shoot electrons through the skin in numbers sufficient to be counted. Such an experiment is fairly safe, because radio-active sodium disappears pretty quickly, whereas radium stays in the body indefinitely.

By such methods as these it has been shown that even the bones and teeth of a living man or animal are constantly exchanging atoms with the blood, and in a more active tissue such as muscle or liver the exchange is extraordinarily quick. Half the nitrogen atoms in some compounds which form part of the living structure of liver may be exchanged in twelve hours. In fact a living man or animal is in many ways more like a flame or a waterfall than an ordinarily solid body. The structure remains, but is constantly being built up from fresh atoms.

Unfortunately for Europe most of this work has been done in the United States. But cyclotrons to produce artificially radio-active atoms will certainly be needed on a big scale in Britain and Europe after the war, both for the treatment of disease, and for research which will influence both medicine and agriculture.

Nature Cures

Our medical correspondent has stirred up a hornet's nest by attacking "nature cures." I don't feel called on to take sides in the controversy, because I don't know of any very strong evidence on one side or the other. I know people who say that their health has been restored by these methods. I know others who say the contrary, and I do not know of any good statistics on the matter.

But I should like to examine one statement of Mr. E. J. Saxon, a supporter of Nature Cures. He says of our medical correspondent "As for the body poisons which he dismisses as 'sales-talk,' there is plenty of objective physico-chemical evidence available on this point. It is largely to the inability of the kidneys and skin to clear these out of the body that a great deal of ill-health and disease is due, and that inability is due to a large extent to the very thing he advocates, eating plenty of 'ordinary food.' "

Now I expect many doctors would agree with him, and so would most of the general public, because they learn their medicine from advertisements. But I don't believe it myself. When the kidneys are diseased in certain ways, or if the excretion of urine is mechanically blocked, as in prostate disease, various substances which are excreted in the urine accumulate in the blood, and if the patient is enabled to excrete them, he generally gets better.

What is more, one can judge the gravity of the disease by the amount of these substances in the blood. Eighteen years ago I was a hospital chemist, and used to determine the quantity of urea in blood. Normally there is about one part in four thousand. If it rose above one per thousand the surgeon would not operate, as the patient would probably die if he did so.

Nevertheless urea, which is one of the main solid constituents of urine, is a pretty harmless stuff. I have eaten four ounces of it at a time with no ill effects. Another famous bogey is uric acid, to eliminate which the unfortunate dupes of advertisements spend hundreds of thousands of pounds a year. In gout it accumulates in some of the joints, and under the skin, and causes a good deal of pain. But in another disease, leukaemia, the blood contains plenty of uric acid, but it does not get deposited in this way. So clearly gout is not due to a failure to get rid of uric acid. Still less is ordinary rheumatism.

Besides this, acid accumulates in the blood in kidney disease, and causes panting. One can acidify one's blood enough to get fairly severe symptoms, and I have done so. But I was certainly not as ill as a case of kidney disease with equally acid blood. The plain fact is that nobody knows what are the poisons which kill people in some types of kidney disease, though there is little doubt that such poisons exist. But I certainly do not know of an "objective physico-chemical evidence" about them. Plenty of people have tried to isolate them, but without success. It is much more doubtful whether poisons are eliminated in the sweat to any marked extent. And it is still more improbable that poisons accumulate in the blood where there is no definite disease of the kidneys. Still less evidence exists that "an ordinary diet" paralyses the kidneys, or the sweat glands. The world's sweating record, of five pounds in an hour, was made nearly twenty years ago by an English coal-miner, and is still, so far as I know, unbeaten. He worked in a very hot mine, and ate a most unnatural diet containing large quantities of bacon and kippers. He needed a lot of salt to make up what he lost by sweating.

The statistics of industrial disease show that the kidneys are badly affected in certain occupations. The Registrar-General gave the mortality in 360 trades for 1930–32. The fifteen with the highest death-rate from kidney disease included innkeepers, barmen, waiters, and bookmakers. This suggests very strongly that alcoholic drinks are bad for the kidneys. Among these fifteen are also cotton blowroom workers, cotton strippers and grinders, wool weavers, wool spinners, and textile trade dyers. Clearly there is some influence in the textile trade which causes kidney disease, though no one knows what it is. If Mr. Saxon thinks that textile workers eat more "ordinary food" than carpenters, locomotive firemen and cleaners, or furniture salesmen, who are among the fifteen occupations with the lowest death-rate from kidney disease, I think he should prove his case. Till he has done so, I think he is diverting our readers from the very real dangers of the drink and textile trades.

My main reason for raising this question is a desire to fight the appalling ignorance on this topic. Advertisers tell the public that back-ache is a symptom of kidney disease. It is a very rare one. Chronic headache and puffy ankles are much commoner symptoms. One firm goes on to swindle the public as follows. They sell pills containing a dye called methylene blue, which gives a blue or green colour to the urine. They tell their clients that the drug causes the elimination of impurities from the body, and certainly the pills do have a visible effect. This sort of thing is legitimate capitalist enterprise, and there is big money in it. Whatever false claims either the medical profession or "unortho-dox" practitioners such as nature curists may make, they certainly do nothing as bad as this.

I have myself little doubt that some people benefit when they stop eating cooked food and in particular some constituents of our ordinary diet. But I don't think everyone would do so. On the contrary, the evidence is very strong that about a third of the population would benefit greatly from more of quite ordinary foods. And I think it is our business, while not forgetting these special cases, to concentrate on the needs of the masses.

Inoculation

The war is leading to a great deal of inoculation. Soldiers are inoculated against typhoid, tetanus, and other diseases. The authorities strongly recommend that children who spend long periods in crowded shelters should be inoculated against diph-theria, and so on. In fact, wherever hygienic conditions are bad, inoculation is put forward to prevent the spread of disease. I am often asked what I think of it. I cannot answer this question, because the word inoculation covers hundreds of different processes, and I do not think either that all are useful, or all dangerous.

Originally it meant infection by way of the eye, and was applied to the practice of giving smallpox in this manner, brought over to England from Turkey some two centuries ago. This was supposed to give a mild attack, but it certainly killed a great many people. The word is now loosely applied to any method by which substances supposed to protect against a disease are intro-duced under the skin.

There are at least five distinct processes. Nowadays people are never given a fully virulent disease, at least not intentionally. They are given mild diseases which protect them to some extent against severer ones. Thus cowpox, or vaccinia, which is used for ordinary vaccination, gives a partial protection against small-pox. And the virus of rabies, weakened by heating or drying, gives some protection against the spread of this disease by the bites of mad dogs. One very real danger of such a method, where an actual infection is transmitted, is that another may be transmitted with it. A number of children have died of inflammation of the brain after vaccination. This may be a rare effect of the vaccinia virus, or it may be due to a separate agent.

Attempts have been made to apply this method to tuberculosis and other diseases, with little success. A weakened form of the tubercle bacillus, called the bacille Calmette-Guérin (B.C.G.) after its inventors, is given by mouth to newborn children to protect them against real tuberculosis. The mortality of children in tuberculous families treated this way is said to have been reduced to one-eighth in France. On the other hand, a number of children at Lubeck, in Germany, were killed when they were given virulent bacilli by mistake. At any rate, very few doctors in this country will use the B.C.G.

A much safer method is to inject dead bacteria. This is done to soldiers to protect them against typhoid fever and other similar water-borne diseases. As the bacteria are dead, and suspended in a mild antiseptic, there is extremely little danger of a real infection developing. But the injection certainly makes one ill. I had to do sentry duty with a fixed bayonet the night after injection in 1914, and to go up to Madrid on the outside of a very ancient lorry the night after injection in 1936, so I know all about it. But I think the evidence is very strong that this injection gives a good deal of temporary protection.

Attempts have been made to apply this method to protection against common colds, and success has been claimed. But when a number of medical students were vaccinated against colds in this way, and an equal number left unvaccinated, no difference was found in the number in the two groups who developed colds. The vaccinated ones mostly said it had benefited them; but if there was any benefit, it was probably psychological, and injections of water and salt would have been as useful, and a lot cheaper.

Still another method is to inject, not bacteria, but a poison produced by them, which has been treated by heat or chemicals so as to be harmless, but still produces immunity. A toxin treated in this way is called a toxoid. This method has been conspicuously successful in protecting children against diphtheria. But it is not likely to be of much immediate use at present,* as the immunity takes about four months to develop. And if we are going to get epidemics of diphtheria in shelters, they will probably be raging long before April.

These three methods produce what is called active immunity. That is to say, they cause changes in the person inoculated similar to those which take place during recovery from an attack of the disease. These always include changes in the living cells, and may include the production of substances in the blood which protect from the disease. When such substances are produced in large amounts, an injection of blood from an immune person will give protection. This is the case with measles. A child which has been exposed to infection can be protected by injecting blood (or more usually serum, the fluid which remains when the rest of the blood has clotted) from a convalescent. This may stop the attack, but generally merely makes it very mild. Unfortunately it is too late to give such an injection when the spots have actually appeared.

This kind of immunity is called passive immunity, and wears off in a few weeks or months. It is only useful in a few diseases, such as diphtheria and tetanus or lockjaw. In these cases the antitoxin, as the immune body in the serum is called, is made by an animal, very often a horse. There is no moral objection to this. The horse is not given diphtheria germs. At first quite a little of the poison produced by these germs is injected, later on, as immunity develops, the dose is many times what would kill a normal horse. But it is never seriously ill, and suffers no more than a blood donor when its blood is drawn. There is, however, a medical objection. The first injection of horse serum is harmless. But a later one may cause an outbreak of rash, or more serious illness. And unfortunately the purification of antitoxins from other substances in animal sera which have these effects is very difficult.

Finally a mixed type of inoculation is used. A mixture of

* December, 1940.

diphtheria toxin and antitoxin may be injected. The antitoxin protects against the harmful effects of the toxin, but in spite of this, immunity develops, though unfortunately only in the course of months. Most of these treatments are pretty safe. But few are absolutely so. Though if the chance of death is one in a million (and it is often less) it is worth taking if we can cut down a chance of one in a hundred of death from some disease to one in a thousand.

Serum is unfortunately a very good medium for the growth of several kinds of bacteria, and although many precautions are taken, it may become infected. Thus at Bundaberg in Australia in 1928, twelve children died of blood poisoning after inoculation against diphtheria.

The proper method of preventing disease is certainly by sanitation, the prevention of overcrowding, and so on. For this reason some people attack all preventive inoculation. Their attitude is like that of the questioners at public meetings on A.R.P. who ask me whether I don't think it would be better to have no wars than to make bombproof shelters. Of course it would. But in a world of capitalist states where war is inevitable, shelters are needed. In the same way, in a city where hundreds of thousands spend the night in grossly overcrowded shelters, inoculation is needed.

Immunization to Diphtheria

We are officially urged to have our children immunized against diphtheria, and last year a number of posters appeared asking "Is immunization safe?" and hinting that it is not. I am in favour of immunization against diphtheria, but I don't pretend it is a hundred per cent safe.

Few things are. Several thousand people are killed each year crossing the street, but we go on taking the risk. What is more, immunization rarely if ever gives complete protection throughout life. Some people die of diphtheria though they have been immunized. Once this is admitted, the main question to be answered is whether one can greatly cut down the risk of dying by being immunized. The opponents of immunization have several other arguments. They point out, quite correctly, that there are other and better ways than immunization of warding off a disease.

A hundred years ago cholera and typhoid killed many people in England. These diseases have been wiped out by the state insisting on a clean water supply. Could not diphtheria, smallpox, and many other infections be dealt with in the same way? The answer is that they could, but only by a campaign covering the whole world, and costing perhaps a tenth as much as the present war. As long as there are areas in India or China where smallpox is constantly present cases will keep on occurring in this country, and some kind of protection is needed against it. This is one of the arguments, though not the most important one, for bringing the standards of living of the whole human race at least up to the not very high level reached in Western Europe before the war. Until this has been done there is no chance of wiping out a great many deadly diseases.

Another argument is that most antisera are made by private firms, which not only make high profits, but often spread exaggerated propaganda to boost their wares. This is also quite true. But it is likewise true of the firms which sell food, coal, clothes, railway tickets, and other necessaries. Socialists can't go about naked till the clothing trade is socialized, and their children should not lack immunization because the supply of antitoxin is largely in private hands. Finally, it is argued that immunization involves cruelty to animals. This in not true so far as diptheria is concerned. There are several methods of immunization against this disease. Diphtheria is due to a local infection, usually in the throat. The bacteria in the throat may cause such a severe swelling that the patient has difficulty in breathing, and may die of croup.

But the main danger is from poisoning by diphtheria toxin, a chemical substance produced by the bacteria. The poison may stop the heart, or cause paralysis of other muscles. Thus the disease is very different from one such as tuberculosis, where the main damage is done in the organ infected with the bacteria, so that the symptoms of lung, bone and skin tuberculosis are quite distinct, and they were long thought to be separate diseases.

The germs of diphtheria can be grown in a variety of mixtures.

They do not need a living man or animal. If their food is properly chosen they make plenty of poison, which can be considerably concentrated. It is then injected into a horse in gradually increasing doses, none of them big enough to produce even the moderate illness caused by antityphoid inoculation. The horse responds by making a substance called antitoxin, which combines with the poison to make an innocuous compound. It is then bled from time to time, and its blood serum used as a source of antitoxin. Except for retired race-horses at stud, these horses have an easier life than any other horses in England.

The antitoxin can be injected into a patient suffering from diphtheria, or one who has been exposed to infection, and greatly reduces the risk of death. But unless one learns to make the antitoxin oneself the immunity only lasts for a few months. Babies of immune mothers are born with enough antitoxin to last some months, but they lose their immunity gradually. A human being can be immunized by gradually increasing doses of toxin, but it is easier and safer to inject a fairly big dose of the toxin-antitoxin compound. This is a loose compound; and breaks up to a slight extent, just as ammonium carbonate breaks up enough to give a smell of ammonia. So when it is injected into a child the child makes antitoxin; and before the injected antitoxin has disappeared, the child has made enough to give it considerable protection against diphtheria.

One reason why the injection of antitoxin is not a hundred per cent safe is that animal serum is a good medium for growing some very dangerous bacteria; and if contaminated serum is injected, a child may be killed. This happened to a number of children at Bundaberg in Australia. Great precautions are taken to avoid it, and the risk is now extremely small. I have read a good many of the arguments on both sides, and am personally convinced that the gain in safety from immunization to diphtheria vastly outweighs the risk, and that adequate precautions are taken to avoid other infections.

I wish I could say the same about vaccination against smallpox. Here the principle is very different. The child is infected with cowpox instead of being injected with a chemical substance. And in practice the chance of its getting another infection as well is far greater than with diphtheria immunization. The vaccinia or cow-pox virus is grown on calves which undergo discomfort, though probably not severe pain. It can be grown in hens' eggs, which certainly cannot suffer, since at the stage of development used, the embryos have no nerves or brains. This method of growing the virus is more expensive, but both cleaner and more humane than the old one, and might well be used. There is another danger, namely that vaccination sometimes, though rarely, causes a fatal infection of the brain, though this usually occurs when it has been put off till a child is some years old, and is extremely rare, if it occurs at all, in young babies.

I believe that our present technique of vaccination is out of date in several respects, and should like to see a real drive to make it safer as soon as possible after the war. I also doubt the efficiency of immunization against common colds. But I am in favour of immunization against diphtheria, and should certainly be immunized myself if a test with diphtheria toxin had not already shown that I am immune.

5

HYGIENE

Overcrowding and Heart Disease

HEART disease is one of the commonest causes of death. But a great many of the deaths from it occur in old age. It does not very much matter whether one dies at 80 or 85. One's work and pleasure are both pretty well over. Death in the early years of life is much more serious.

One of the great causes of heart disease in children and young people is rheumatic fever, or acute rheumatism. It has long been known to be a disease of poverty. The fact that several cases often occur in one family has been put down both to infection and to heredity. But poverty is the main cause. How it acts is another question. Some have blamed bad feeding, others bad clothing, damp, overcrowding, vermin, and so on. As all these are consequences of poverty, it was hard to discover which was most to blame.

Some progress has been made by a very important research suggested by Professor Perry, of Bristol University, and carried out by Dr. G. H. Daniel. He investigated 341 working class families in Bristol containing one or more children with rheumatic heart disease. At the same time the University of Bristol and the Colston Research Society were making a social survey, published in 1938, and enough families were studied to make it possible to say what proportion of families with a particular standard of living were affected with this form of heart disease.

The families were not classed by their total income, for clearly a family with one child and £3 per week is better off than a family with six children and £4 per week. They were assessed on the relation between the family's total income and its minimum needs according to a standard laid down by Mr. R. F. George in the *Journal of the Royal Statistical Society*. The

results were clear enough. Seventeen per cent of the families
fell below Mr. George's poverty line, and among them the
frequency of rheumatic heart disease was 39 per cent above the
average. In the 22 per cent who had double his minimum the
frequency of heart disease was 23 per cent below the average.
It is quite clear from the figures that in order to cut down the
rate of heart disease in children below the working class average,
an income 50 per cent above Mr. George's standard is necessary.
Once this is reached, further additions are not very important.
Judged by objective standards like this, a great many minima
proposed by economists would be found to be much too low.

Heart disease in children was still more strongly influenced by
overcrowding. In families with less than 0·6 rooms per person
the disease was 67 per cent above the average. In those with
1·8 rooms or more per person it was less than half. Overcrowding
is a more important factor than mere poverty. However, Dr.
Daniel showed quite clearly that poverty does not act only
through overcrowding, and that overcrowding is dangerous
apart from poverty. Children's heart disease was still above the
average in families with twice Mr. George's minimum income so
long as they had less than one room for each member of the family.

Several other characteristics of the families were recorded.
Families living in a basement have long been known to have
more rheumatic heart disease than others. This was proved to be
so, even when allowance was made for low wages and overcrowding, though of course the effect was less. For very few
people live in a basement if they can afford to live upstairs.
There was a distinct advantage in belonging to a doctor's club,
but none in getting school meals, which suggests that malnutrition is not an important cause of this particular disease, though
it is already known to contribute to a great many others.

Here is Dr. Daniel's summary of some of his most important
findings in his own words: "Thus if the standards of the 30 per
cent of the Bristol working class population with the most
inadequate incomes and housing accommodation were raised to
the average level of the rest of the working class population, a
decrease of 26 per cent in the number of cases of rheumatic
heart disease could be expected. And if standards were raised to

the level of the highest 10 per cent of all working class families, the incidence of the disease would be roughly halved."

It is worth remembering that even if the Beveridge Report is adopted in full, it will only bring the poorest workers up to the average, if that. The Beveridge standard would be an immense improvement, but it would not give all that is needed in the case of this particular disease, or a great many others. On grounds of national health the Communist Party is fully justified in asking for more. And Mr. Bevin's two million builders will all be needed to bring us up to the standard of about two rooms per person which Dr. Daniel's work shows is needed to cope with this particular disease.

I mention this work in such detail because we know much less about the effects on health of overcrowding than of malnutrition. Naturally the two go together. Statistical work of this kind is needed in a score of towns and rural districts, and on many different diseases, before we have a scientific basis for a national housing policy, as we have for our national food policy. We do not know whether there is a gain in health from two small rooms rather than one big one, how important sunlight is in bedrooms, what is the healthiest form of heating, or many other important things. But we do know that our present housing standards condemn thousands of children to crippled lives and early deaths.

Overcrowding and Children's Diseases

It is a scientifically proved fact that in our towns poverty is responsible for a large proportion of the deaths. As we go down the economic scale the death rate from almost all diseases goes up. The exceptions to the rule, such as gastric ulcer, diabetes, and liver cirrhosis, are favoured by overeating or overdrinking. But it is very important to know how poverty acts. In the last twenty years we have at last got scientific standards for diet, and we know that malnutrition is responsible for a great many deaths of babies and of women in childbirth, besides lowering resistance to many diseases.

Scientific studies on the effects of bad housing are much more difficult. Among the most striking are those of Professor Payling

Wright, of Guy's Hospital, and his wife, on the death-rates of young London children from diphtheria, measles, tubercle, and whooping cough in the years before the war, published in the *Journal of Hygiene* for 1942. On an average 660 London children under five were registered as having died of these diseases each year. But another 1,100 died of bronchitis and pneumonia, and as these deaths went up with each epidemic of measles or whooping cough, the true mortality is probably over 1,000.

The Wrights compared the death-rates in the twenty-eight London boroughs, with East Ham and West Ham, and related them to social conditions in 1931. The economic levels of the boroughs were estimated from the proportion of the population below the "poverty line" set by the *New Survey of London Life and Labour*, and from the average income left over after rent had been paid. The overcrowding was measured by the percentage living two or more per room. Naturally low wages and overcrowding went together as a whole, and there were big variations between the boroughs. Thus only 15 per cent of the adult males of Hampstead, compared with 55 per cent in Bermondsey, were in the worst paid occupations.

Nevertheless the differences suffice to answer the important question, "Is poverty or overcrowding more important in killing children?" If we make a graph in which the distance of a point to the right of one line represents the percentage of overcrowding, and the distance above another gives the death-rate from measles, we get a point for each borough. These points lie in an irregular cluster sloping up and to the right from Hampstead to Bermondsey. We can calculate a number called the coefficient of correlation, which would be equal to nought if overcrowding had no effect on measles, and to one if nothing else had any effect, and all the points lay on a straight line.

We can do the same for poverty and measles, and in each case find a high correlation. But a competent statistician can do more. He can calculate what is called a coefficient of partial correlation between measles and poverty, with overcrowding held constant. That is to say he can ask, "In boroughs with the same standard of overcrowding, do more children die of measles in those which also have a low average wage rate?"

The answer is quite clear that they do not. Poverty has a great effect on the death rate from measles, but it acts entirely through bad housing. The same is true for whooping cough. On the other hand poverty has a moderate effect on deaths from diphtheria, and a very big one on deaths from tubercle, apart from the effect due to overcrowding. Poverty probably helps the tubercle bacillus through undernourishment and bad clothing and heating, as well as bad housing. In other words, if the families of the poorer boroughs were housed as well as those of Hampstead, at their present rents, but without any rise of wages, the death-rates of children from measles and whooping cough would probably fall to the low level of Hampstead, deaths from diphtheria would fall most of the way, but those from tubercle only a little. As I hope to show in a later article, bad housing, as well as low wages, plays a rather big part in causing adult tuberculosis.

It is easy to see the main reason why overcrowding makes measles and whooping cough so deadly. Most children get them some time, and the parents are not unduly alarmed. But measles, and particularly whooping cough, are deadly diseases in babies. A child in its first year is about seventy times as likely to die if it gets whooping cough as one in its fifth, and about fifteen times as likely to die of measles. The difference is not so large for diphtheria. The more overcrowded is a borough, the earlier the children get infected on an average, whereas poverty by itself has little effect on the age of infection. If any child in an over-crowded house gets one of these diseases at school they are almost certain to infect their baby brother or sister and may kill them. We cannot yet lay down a standard for housing as scientific as our standards of food. But we can say that any family where the baby is necessarily exposed to infection by its elder brothers or sisters is certainly overcrowded, and can at once condemn any housing scheme which does not make the segregation of the baby possible.

Valuable as the Wrights' work is, there is a great deal more to do on the same lines. Is overcrowding in the schoolroom as dangerous as in the home? Is a damp, cold, and sunless house as bad as an overcrowded one? Do playgrounds near the home make a difference? Such are some of the obvious questions to be answered. Nevertheless such work is of the utmost use, both in proving the very great importance of housing for our national health, and in showing that while decent housing would save many children's lives, it is not a cure-all.

Overcrowding and Tuberculosis

The fall in the death rate from some diseases is due to special hygienic measures based on scientific knowledge. Thus cholera has been abolished and typhoid made very rare in Britain by giving the towns a proper water supply, filtered or chlorinated so as to remove the causal bacteria. But in other cases the improvement is almost wholly due to improved economic conditions. This is well brought out in Hart and Wright's recent book on *Tuberculosis and Social Conditions in England*, published by the National Association for the Prevention of Tuberculosis.

The death-rate from respiratory tuberculosis (mainly phthisis) has been falling for 100 years. In 1835 about 400 men in every 100,000 between 25 and 45 died of it, in 1935 about 100 did so. The one exception to this fall was found among young women between 15 and 35. Their death-rate fell steeply until 1900, but the fall was then checked.

The graph shows the death-rate among girls and women between 15 and 24. It is plotted logarithmically, which means that a fall from 200 to 100 per 100,000 per year is represented as large as a fall from 100 to 50. So if the graph were a straight line, it would mean that in every 35 years, or some such period, the death-rate was halved. The other curve shows the average real earnings of a worker, allowing for unemployment, changes in prices, and social services. These are based on figures of bourgeois economists such as Bowley and Wood, and on the figures of the Ministry of Labour, which if anything paint too bright a picture of recent years.

A glance shows that one curve is almost identical with the other turned upside down, except that the sharp changes in the death-rate come a year or two after those in the standard of living. This is natural enough. Tuberculosis kills fairly slowly, so there

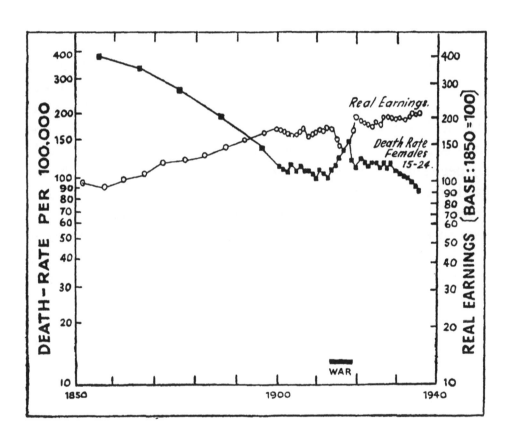

is bound to be a lag between economic cause and hygienic effect. But there can be no reasonable doubt that tuberculosis in girls is far more closely dependent on economic causes than tuberculosis in older women, or in men not liable to lung injury from dust.

In order to get further information on how poverty acts, the figures from 76 English county boroughs were analysed. They were classified in several ways, and particularly by the percentage of the population receiving poor relief in 1931–33, which varied between 9·5 per cent in Sheffield and 8·4 per cent in Lincoln to 0·9 per cent in Halifax and 0·7 per cent in Oxford. But this showed only a moderately close relation with tuberculosis.

The picture was very different when the boroughs were classified by the percentage of people living at a density of more than two per room. The five worst boroughs, Sunderland, with 29 per cent, Gateshead, South Shields, Tynemouth, and New-castle-on-Tyne with 23 per cent, are all in the north-east coast areas. They compare with 2 per cent in Northampton, and 1½ per cent in Bournemouth. In the boroughs where over 10 per cent had less than half a room apiece the death-rate among girls and women between 15 and 24 was actually higher in 1932 than in 1912, and that among boys and men of the same age had barely fallen. The differences in other age groups were not so striking. Any fall in the standard of living, such as occurred in the 1914–18 war, and in 1921, immediately started killing the girls of the congested areas.

It is quite obvious that, whatever can be done for tuberculosis at other ages by sanatorium treatment, improved feeding and so on, we cannot hope to do anything serious against phthisis in young women, or much against it in young men, except by abolishing overcrowding. The death rate in girls began to fall about 1933, probably as the result of the great housing schemes started by the second Labour government.

It is important to note that overcrowding is not such an important factor in causing the majority of diseases as it is with phthisis in young people. It has a big effect in putting up deaths from measles, whooping cough, and rheumatic heart disease, as I have shown in earlier articles; but Hart and Wright point out that for other diseases taken as a whole, other factors in poverty are quite as important as overcrowding. To take an example from another lung disease, Meakins and McKenna examined 200 cases of lobar pneumonia treated with sulphonamide in the Royal Victoria Hospital at Montreal. Twenty-one of them died, and 11 of these 21 had signs of malnutrition. The higher death-rate from pneumonia among the poor is probably much more due to malnutrition than to overcrowding. And McGonigle showed that slum clearance may even raise the death-rate if it means such a rise in rent that the tenants have to cut down their food purchases. It is quite clear that the fight against poverty must be fought on several fronts. The main front is the attack on low wages, but a gigantic building programme is a second front without which the other cannot hope for full success.

It is most encouraging for the future of medical science that a professor of pathology in one of our great medical schools should take up research of this kind. But a vast amount remains to be done. We know very little about the effect of overcrowding at work and in vehicles. You may add several years to your life by cycling to work rather than going in a crowded tube. No one knows. But this work brings the day nearer when we shall have scientific standards of housing comparable with the scientific standards of diet which are the basis of our rationing system, and have done so much to keep us reasonably healthy during the war.

Dangerous Trades

Among many excellent proposals in the Beveridge Report there is one which, if it is carried into effect, will lead to great advances in industrial health. Industries which show an abnormally high death-rate and sickness-rate are to be specially scheduled, and two-thirds of the extra cost of compensation of workers in them is to come out of their profits.

Many people may think that this is already done under the Workmen's Compensation Acts, which compel the employers to pay the victims of accidents, and of a few industrial diseases. But actually the large majority of the casualties of industry are not compensated at all.

Take the example of the pottery trade. Potters have a death-rate 35 per cent above the average. Some of the excess deaths are due to silicosis, and compensation ought to be given in such cases. But, in addition, their death rates from phthisis and bronchitis are more than double the average. In any particular case one cannot be sure that the potter would not have died of the disease if he had worked in a brick kiln and not a pottery kiln. But his death was probably due to injury of the lungs by silica dust. If the Beveridge Report goes through, much of the cost of compensating these men will be thrown on the industry. This will give the capitalists of the pottery trade a real economic incentive to improve its health conditions. The pottery research institute at Stoke will begin to interest itself in broken potters as well as broken pots. And it may be that lung disease among potters will be drastically reduced, as lead poisoning was when its victims were compensated.

What are the industries which will probably be scheduled as hazardous? If the Registrar-General's Report on Occupational Mortality published in 1938 is taken as the measure of industrial hazard, there is little doubt that one of the most dangerous of the large trades is the drink trade, which employed about 110,000 men, whose death-rates ranged from 55 per cent above the average in the case of innkeepers, to 16 per cent above in the case of brewery workers. Undoubtedly the excess deaths are largely due to drinking too much. At present if a brewery owns a public house it pays them that the innkeeper should drink as much as possible. If it became worth their while to lower the publicans' death-rate this might not be so.

But the biggest blot on the industrial health map is undoubtedly water transport. Railwaymen are conspicuously healthy, and transport workers near the average; but merchant seamen, dock workers, and even bargemen, have a death-rate far above the normal. Seamen die over twice as often from violent deaths as other men. It is often said that this is inevitable, as the sea is more dangerous than the land. But fishermen, who are healthier than the average, have a death-rate from accidents which is only 6 per cent above that of the rest of us. One reason for the difference is certainly that fishermen control their conditions of work to a far greater extent than seamen. Apart from accidents, seamen have a very high death-rate from tuberculosis, and a fairly high one from diseases associated with alcoholism. Both these would be lowered if they had better quarters and better ventilation on board ship, and better alternatives to the pub when they went ashore. If the Beveridge Report goes through, the ship owners will have an economic incentive to provide both. Stevedores have not only a terrific accident mortality, but die of a great many diseases probably due to bad housing conditions. Out of two hundred occupations, only three have higher death-rates.

The other excessively hazardous trades are mostly branches of larger industries. Thus the death-rate among coalminers as a whole is only 5 per cent above the average. But the anthracite miners in South Wales have a mortality 43 per cent above the average. This is largely from silicosis, but also from accidents.

Again, the glass trade as a whole is not very unhealthy, but glass blowers, who form only a sixth of all glass workers, have a death-rate of 60 per cent above the average. Similarly in the textile industry a few dusty trades, such as those of strippers and grinders, and blow-room workers, are far more dangerous than the remainder.

The building trade as a whole is decidedly healthy, but masons working in sandstone have a death-rate of 80 per cent above the average, with silicosis 55 times the average; and the quarrymen who produce the sandstone are also very liable to lung diseases.

If it is possible, for purposes of taxation, to separate these particular occupations, it will certainly make for health. It will mean, for example, that the price of sandstone will go up. And a good thing too. As long as sandstone is dearer in human life than limestone or brick, it should be dearer in money too. No doubt the capitalists in these dangerous trades will raise the same bitter cry as the insurance magnates. They will say that they take every possible precaution, and that the high death-rate is the workers' fault, or inevitable for some reason or other. Doubtless the best cure for such ideas would be to make a few shipping magnates sleep in the foc's'le for a year or so, and put colliery directors to hew anthracite.

Until that can be done, I hope that the whole Labour Movement will see that the capitalists in those industries which suck the life-blood of the nation are made to compensate the workers whose healths they have ruined and the widows whose breadwinners they have killed.

But the Beveridge Plan will be unworkable if we have unemployment after this war on anything like the scale that we had after the last. At Oxford on December 6, 1943, Sir William Beveridge said of the abolition of mass unemployment, "I do not know how it is to be done, and I do not know whether anybody else knows." It is a pity that he has not heard of Stalin.

NOTE.—This proposal for the special taxation of dangerous industries is one of the features of the original Beveridge Report which the Government rejects in its White Paper of 1944.

The Drink Trade

My recent article on dangerous trades has brought a reply from a Mr. Scott London, who does not like what I wrote about the Drink Trade. He hopes that the *Daily Worker* will publish it. Unfortunately it is a little longer than my own article, and there might be a bit of a dust-up on the Editorial Board if I tried to substitute it for one of my own. But here is an offer to you, Mr. London.

If you will induce those distinguished representatives of the brewing trade, Lords Iveagh and Moyne, for whom Guinness is good, whatever it may be for others, and Colonel Gretton, M.P., of Bass, Ratcliff, and Gretton, to press the Government for those extra tons of paper for us, we will print your article when we get them.

Till then, I can only answer some of Mr. London's points. "The professor deduces," he writes, "without evidence, that the excess deaths are largely due to drinking too much." Very few deaths are registered as due to alcoholism, but when we come to diseases largely caused by alcoholism, the case is quite clear.

One of these diseases is cirrhosis of the liver. This is not a very common disease. Round 1931 it only killed 1,028 men and a good deal fewer women each year. But it killed 93 innkeepers per year, and an innkeeper was over eleven times as likely to die of it as an ordinary man.

It is quite an interesting disease from the economic point of view, because it is one of the small group which kill the rich more than the poor. Among the best off $2\frac{1}{2}$ per cent of men the death-rate from cirrhosis was 84 per cent above the average. Among the middle class it was 95 per cent up. But this, in the words of the Registrar-General, "was due mainly to the high rates among innkeepers and their wives, proprietors of businesses, commercial travellers, and employers and managers in certain industries."

The skilled workers showed the lowest death-rate of any classes, but the semi-skilled and unskilled workers were also below the average. Of the deaths actually registered as due to alcoholism more than half occurred in the rich and middle classes, though they include only 16 per cent of the population. That is part of the answer to those who say that drink is the curse of the working class.

The other occupations with a specially high death-rate from cirrhosis, though all below innkeepers, are, in order, actors, barmen, bookmakers, lawyers, musicians, wholesale business proprietors, retail proprietors (dairy, meat, fish and greengrocery), bank and insurance officials and doctors. No railway signalmen and no compositors died of it in 1930–32.

The inn- and hotel-keepers also had the highest death-rate of 198 occupations from diabetes and from diseases of the digestive system. They come second on the list for kidney disease, and third on that for "cerebral vascular lesions," in ordinary language, stroke. They had the second highest suicide rate of eighty-six occupations. It would be interesting to know to what agency other than drink Mr. London attributes these facts.

He goes on to write, "In the total of 110,000 men for the drink trades, it is interesting to assess how many are innkeepers with the high death-rate of 55 per cent above the average, and how many brewery workers with the lower rate of 15 per cent. At a guess about half of 1 per cent, which gives a proportional figure of about 16 per cent for the trade."

Mr. London's guess is that there is one innkeeper to 199

brewery workers. Actually in 1931 there were 17,033 male "makers of alcoholic drinks" in England and Wales, and 65,183 innkeepers. There were also 24,309 barmen, with a death-rate of 49 per cent above the average. But Mr. London, who states that he "champions the truth," prefers his guess to the census figures.

Mr. London next tries to explain the high mortality among innkeepers from the fact that they take up their job late in life. This fact is of course allowed for by the Registrar-General, who is not a complete fool. If he had not allowed for it, their death-rate would have appeared to be 174 per cent above the average.

Mr. London goes on to air his remarkable views on statistics. He says that, "incidence of mortality, or expectation of life is a probability, no more subject to scientific laws than is the chance of birth." In fact a whole branch of science, of which I am a professor, deals with such matters.

No quantitative science reaches absolute certainty. It is fairly easy to measure an inch accurately to one part in a thousand. It is just possible to measure it to one part in a million. It is impossible now, and perhaps always will be, to measure it to one part in a thousand million.

We cannot often say that a particular man will die next year, but, apart from war and epidemics, we can say with great certainty that between three thousand and four thousand of a particular group will do so.

The certainty is still greater in chemistry and physics. No one can say that a particular molecule of water in a boiler will go off in steam in the next minute. But we can be sure within less than 1 per cent what fraction of all water molecules in a boiler will do so. If we had not this kind of certainty, most branches of science would be impossible.

But it is not a sheer accident that defenders of the drink trade prefer unsupported assertions such as "— is good for you" (no prize offered for filling in the blank) rather than a study of statistics. You do not have to be a very advanced Marxist to explain that social phenomenon.

Factory Ventilation, Heating, and Lighting

Joint production committees are doing a great job all over the country, not only in increasing production, but in improving conditions of work, so that a worker can produce more with no greater effort, or even with lessened fatigue. To do this efficiently they need scientific information, and particularly methods of measuring the conditions in their factory. They will find very useful materials in the Industrial Health Research Board's 3d. pamphlet on *Ventilation and Heating, Lighting, and Seeing*, published by the Stationery Office (postage extra). It is not perfect, but is very good value for the money.

Ventilation should be at a minimum rate of 17 cubic feet of fresh air per worker per minute, and a good deal more in summer, or when the air contains harmful dust or fumes. It is rather hard to measure this rate. But it is possible to measure the rate of air movement with an anemometer, and every big factory should have one. Even in winter this should not fall below 20 feet a minute.

It is easier to measure temperature. How many shop stewards know that under the 1937 Factories Act, the temperature in rooms where a substantial proportion of work is done sitting, and does not involve serious physical effort, must not be below 60° after the first hour's work? Rather lower temperatures are desirable when work is hard.

How important this can be is shown by the record of accidents in munition factories in 1914–18. They were at a minimum with a temperature of 65–70° F., and went up by 20 per cent when the temperature rose above 75°, and by 30 per cent when it fell below 55°. So a thermometer in your factory is a good investment for the insurance company as well as for the workers!

Of course it is difficult to keep temperatures down in summer, particularly when the black-out interferes with ventilation. A working rule suggested is 5 square feet of ventilation opening for every 100 feet of floor space. Does your factory reach this standard? If it does not reach it at night, a great deal can be done by light traps at the windows, and fans where necessary. The incoming air can be heated in winter by passing it over a steam radiator. Such unit heaters are often used. By connecting the intake side to the outer air by a duct which is shut in cold weather, they can be made to cool the factory in summer as well as heating it in winter.

Good lighting is just as important for health and efficiency as good heating. The amount of light needed varies very much. Fine work needs over one hundred times as much light as the roughest work. However, a minimum is laid down by the Factories (Standards of Lighting) Regulations (S.R. and O. 1941, No. 94). For most work this is 6 foot-candles at bench level, or 3 feet above the floor. This is the intensity of illumination given by six standard candles 1 foot away, or a 600 candle-power lamp 10 feet away, and so on.

Illumination is quite easy to measure with a light meter, which compares the light in a factory with a standard. "This instrument," says the report, "can be obtained in a very small and convenient size for factory use . . . and should be widely known and used, in order to keep a check on the illumination available, and to encourage proper maintenance of lighting installations." If you want to encourage your boss, and can't get hold of a light meter, you will very likely be able to borrow one from the local electric light company. For fine work a lot more light than 6 foot-candles is needed, and the regulations demand that it should be supplied.

Besides lighting on the job, general lighting is very important. If the factory consists of alternate patches of glare and gloom, movement and cleaning become difficult, accidents are increased, and the workers are more easily fatigued. The best effects are obtained with very modern methods, such as fluorescent lamps. But even where the lighting cannot be increased, one can often get a conspicuous increase in brightness by painting the walls in light colours, and keeping lamps and reflectors clean. If the boss thinks you have all the light you need, and calls you a squander-bug if you ask for more, here is what the Report says of him. "The cost of lighting—small as it is—leads some people who have to provide it to persuade themselves that poor lighting is better than it really is. This is being penny wise and pound foolish; for, in the long run, bad factory lighting never pays."

The last section of the pamphlet is about spectacles. Many people with normal vision need them for fine work. They can do it without them. But it is an effort to focus on a near object, and this either leads to eye strain or to pausing to rest the eyes. Others, especially those who do not read much, may have quite serious eye defects without knowing it. Does your factory Medical Officer test the vision of all new workers? And can anyone who complains of eye strain or headache have his or her vision tested at the factory? If not, it is pretty sure that some workers, who have no idea that spectacles would help them, are straining their eyes, wasting their energy, and perhaps endangering their lives and those of their mates.

The biggest defect in this Report is that, except in a few cases, it does not state what are the legal rights of the workers to air, heat, and light. Fortunately the Labour Research Department can generally provide this information. I suppose it is too much to hope for it in an official publication of a capitalist government, even though it would speed up production and help to win the war. In spite of this serious omission, I think many shop stewards committees would find the Report in question a good threepence worth.

Badly Housed Occupations

As professor of biometry, I am concerned with the application of statistical methods to biological problems, and I naturally look at social problems from the same point of view. I am not such a fool as to think that the study of human society is merely a branch of biology. Man is an animal, but differs from other animals in being able to plan his own future and that of the community of which he is a member. If the people who regard history and politics as merely human biology were logical, they would also regard biology as merely the physics and chemistry of animals, and say that there is no special science of life.

But the converse error is even worse. Man cannot act as a social being adequately unless he is healthy, and not at all unless he is alive. And those who dismiss the biological approach as materialistic, and say we should keep our minds on higher things, generally turn out to be comfortably off themselves, and preach contentment to those who are not.

What can a biologist say to the demands of the miners and agricultural labourers for better pay and housing? First let us see how a case could be stated against them, based on biological facts. One would say that the agricultural labourers were already longer lived than most people, the mortality being only 71 per cent of the average, and actually slightly below that of farmers. Thus it could be argued that there was no need to improve their conditions. Similarly the coalminers have a death-rate only about 10 per cent above the average. This is partly due to accidents, partly to the big death-rate of anthracite miners from silicosis. As far as deaths from disease are concerned, the coalminers are better off than the majority.

The best answer to such an argument can be got from considering the mortality of their wives and children. The Registrar-General has only recently begun to classify deaths of married women by their husbands' jobs, and the only available figures refer to the years 1930, 1931, and 1932. The farm labourers' wives had a mortality 88 per cent of the average, and their children under one year of 96 per cent. This means that while the men get the benefit of a healthy outdoor life, their wives and children had to bear the brunt of bad housing conditions, and died sooner than the wives and children of the well-to-do class in towns, or those of men in some of the healthier urban trades such as carpentry.

The miners' wives present a very black picture indeed. The wives of coal hewers had 40 per cent more deaths than an average sample of married women of the same age, and those of other groups of coal miners were above the average. Their children had an extra mortality of between 30 and 40 per cent. No one who has visited a number of mining villages can have much doubt why this is so. In spite of a number of exceptions, the housing conditions are generally very bad, even without the effects of subsidence. It is extraordinary to come across a typical urban slum in the middle of a Scottish moor. And in 1930–32 the miners were suffering acutely both from unemployment and from the wage cut which provoked the General Strike.

The only other large trade showing as bad figures as the miners for mortality of wives and children was that of dockers, though horse transport ran it close. Where the good or bad health of an occupation is due to its special conditions, the wives may be reasonably healthy. Thus potters have a mortality 35 per cent above the average on account of the silica dust which they inhale. But their wives are only 2 per cent above the average and their children 2 per cent below. The one marked exception to this rule proves it. The wives of publicans have a high mortality, largely from diseases due to alcohol, because they share their husbands' exposure to alcohol, while potters' wives do not share their husbands' exposure to dust.

When the wives and children have a high mortality, this means that the husbands share the bad home conditions, at least during the night; and a high mortality among wives and children generally goes with a high mortality among their husbands, though the converse is not true. What does this mean in practice? It means that although a thousand miners are killed every year in accidents, more miners' lives are likely to be saved by better pay and housing than by increased safety measures underground, important as these are. Much the same is true for dockers and some other transport workers.

On the other hand, even though neither wages nor housing are as good as they should be in Stoke and Burton, it would probably be easier to save lives by improving the working conditions of potters and brewers than by raising their pay or giving them better homes.

To turn to the brighter side of the picture, teachers probably have the best all-round record for a healthy family life, though the wives of nonconformist clergymen are equally healthy, and those of bank officials somewhat more so; and the children of doctors and Anglican clergymen have a lower death rate.

These occupations show what could be done. We should aim at a society where every baby has as good a chance of survival as a doctor's or parson's, and every wife expects as long a life as a teacher's or bank official's.

When Air Burns

I was badly caught out by a questioner at a meeting of the Socialist Medical Association which I addressed in Glasgow in January on Industrial Health. I did not know that two electric welders had recently been killed by nitrous fumes, though I knew that the health of such workers was endangered.

I have two excuses. One is that I was talking about the major causes of death. And even one or two deaths per year from this cause are almost negligible compared with the hundreds of deaths from silicosis, mostly registered as pulmonary tuberculosis, and not even compensated. The other is that I cannot keep up with medical journals if I am working in a factory.

The danger to welders is from nitrogen peroxide. This is a brown gas, or more accurately a vapour, for it forms a liquid which boils at about 72° F. In the gaseous state it is generally called "nitrous fumes," since it looks like a smoke, although it does not consist of solid or liquid particles, as smokes do. Many people know this gas, because it is given off when nitric acid attacks a metal. But in electric welding it is formed from the gases of the air. Roughly speaking, one-fifth of the air consists of oxygen, and four-fifths of nitrogen.

For a great many purposes the nitrogen might as well not be there. An airman at 40,000 feet is under a pressure of a fifth of an atmosphere. But if he breathes pure oxygen, he gets nearly as much per breath as at ground level when breathing air.

All elements but the inert gases, such as neon and argon, can be made to combine with oxygen, and most of them can be made to yield energy while doing so. This fact is used in ordinary engines, and it would be theoretically possible, though expensive, to run an engine by burning iron filings in oxygen. But nitrogen will not burn spontaneously, mainly because the two nitrogen atoms in a molecule are very firmly bound together. It is necessary to add energy in order to make nitrogen combine with oxygen. Once formed, the various oxides are fairly stable. But they all belong to the class of bodies which chemists call metastable, meaning that in the long run they will turn into something else, and that there is generally a way of making them do so quickly. The best known metastable compounds are explosives, but the change may be a very gentle one, such as a change of crystalline form.

Nitrogen peroxide is nearly as poisonous as chlorine, but its action as a poison is more like that of phosgene. A man gets a breath or two of it, coughs violently for a few minutes, feels better, and walks home. Within a few hours the lungs begin oozing fluid like a bit of sore skin, and if this occurs on a large enough scale he cannot absorb oxygen, and is suffocated within a day or two. If smaller amounts are breathed over a long time, nitrous fumes can cause chronic bronchitis and sore throat. They may also rot the teeth, and have been said to affect the heart and nervous system.

Workers with nitric acid are generally given some protection from them, but electric welders often seem to be neglected, though they are in real danger if working inside a compartment of a ship. There are two forms of protection. The gas may be removed by suction. Or if compressed air is available the workers may wear helmets with a constant supply of compressed air. Fortunately it is easy to test for the presence of this gas. It turns damp blue litmus paper red, as do acid fumes of all kinds. And if you smear a strip of paper with boiled starch and a solution of potassium iodide, it turns blue, especially if moistened. This is a rather more specific test, though chlorine and some other gases have the same effect.

These chemical tests are better than detecting nitrous fumes by their smell, since the sense of smell is very easily fatigued. One may smell a gas on entering a room or compartment, but fail to notice it five minutes later, though there is as much there as before.

The history of industrial health shows quite clearly that the workers are unlikely to be protected unless their unions take the matter up. This is particularly the case with chronic poisoning; for death or acute illness cannot be pooh-poohed, but mild bronchitis or tooth trouble can be.

The nitrogen of the air becomes valuable once it has combined with oxygen, or, in ordinary language, been burned. One way of making these gases combine is by passing air through electric arcs, and catching the nitrous fumes in water, where they form nitric acid. Thus the same process which kills electric welders gives a valuable constituent of explosives and fertilizers. This process is used where water power is plentiful but coal scarce, for example in Norway and in the northern parts of the Soviet Union. A little nitrogen is combined in this way during thunderstorms, and helps to keep the soil fertile. But more nitrogen is fixed by bacteria, especially those which live in nodules on the roots of peas, beans, vetches, and other similar plants. Unfortunately no one has yet found out how the bacteria do it, or we could imitate them, and probably fix nitrogen more cheaply than with an electric arc.

All this is elementary chemistry, and no doubt many of my readers know it better than I do. But chemistry will not be fully used until the workers know the chemistry of their own jobs, and are able to insist on the measures needed to protect their health.

X-rays and Their Dangers

Most of the questions which readers ask me to answer in letters are impossible, if only because they do not give me enough information. Sometimes I can give a partial answer. Mr. Willett asks me about the dangers to which his daughter, and other women working in the X-ray department of a hospital, are subjected.

X-rays are of the same general character as rays of light and heat, or beams of radio waves. That is to say they are best regarded as trains of waves in the "ether" moving at three hundred thousand kilometres per second. They are started by the blows of a stream of high-speed electrons on a metal target in a vacuum tube. Their best-known properties are that they can go through solids opaque to ordinary light, and can then be photographed. As bones are more opaque to them than flesh, they are particularly useful for photographing bones which are abnormal, but they also reveal splinters of metal or glass, tuberculous lungs, and many other diseased organs.

Unfortunately they have another property which is not so advantageous. A man consists of many millions of millions of cells, some of them drawn out into nerve and muscle fibres, but mostly too small to be seen without a microscope. In their ordinary state, cells are not very sensitive to X-rays. But they are very easily damaged when dividing. Now growth occurs largely through the division of cells. On an average, the cells in our bodies are separated from the single cell from which we start by about fifty divisions. The first gives two cells, the second four, the third eight, and so on. Anyone who enjoys multiplication can calculate how many there are after fifty divisions.

This at once gives us a clue to the places where X-rays are likely to hurt us. In an adult the cells in the muscles, nervous system, and many other organs do not divide. But those in the lower layer of the skin do so. For the skin consists of cells which are constantly dying and being renewed from below. In the same way there is wear and tear of the lining of our stomach and intestines, and there, too, a layer of cells is constantly dividing. The cells in our blood also wear out in the course of a month or so, and new blood cells are being constantly made in the marrows of our bones.

Besides this the cells in our gonads (ovaries or testicles) are constantly dividing to form eggs so small as to be barely visible, and spermatozoa much too small to be visible without a microscope. These may unite to form a new individual. Finally a cancer, which grows very rapidly, is for this reason susceptible to X-rays, and it is sometimes, but unfortunately not always, possible to kill a cancer with X-rays, without killing the organs round it. So much for adults. But in a rapidly growing child cells are occasionally dividing in most parts of the body, and in an unborn baby they are dividing everywhere. We can now see which are the danger points for X-ray workers.

X-rays with long waves, made by discharges at a comparatively low voltage, have little penetrating power, and are stopped by the skin. Many of the early X-ray workers suffered from this. They developed "burns" on their skin, which, unlike ordinary burns, did not heal, and sometimes became cancerous. Many of them lost their hands.

Modern X-ray apparatus produce much "harder," that is to say more penetrating, X-rays, and serious injury to the skin is rare. Nevertheless precautions should be taken against this risk. Probably the most sensitive organ is the bone marrow. It is easy to detect early stages of injury to this by examining a drop of blood under the microscope. When Rutherford was in charge of the Cavendish physical laboratory at Cambridge, where there were not only X-ray apparatus, but radium, which produces similar effects, the blood of all workers was examined several times yearly. This ought to be done as a routine where workers are exposed to X-rays over any long period.

If a pregnant woman is X-rayed the result may be a miscarriage or the birth of a severely deformed child. But this does not happen with the relatively small dose needed for a single photograph, which is well worth taking if twins are expected.

A heavy dose of X-rays on the gonads will produce sterility, which usually passes off. But experiments on animals and plants show that even a fairly light dose may produce changes in the offspring. These may be visible or invisible in the first generation. But even where they are invisible at first, which is often the case, they may appear in later generations. These changes are very rarely beneficial, but generally harmful, even if they are useful to human breeders. For example, Soviet biologists in many parts of the Union have to breed parasitic insects which attack the moth *Sitotroga cerealella*, whose caterpillar is a pest of granaries. This is of course done in the case of hundreds of insects all over the world, for example with the common white fly which infests English tomato houses.

But the Soviet workers did not want their moths to escape and start laying eggs in neighbouring grain stores. So Volkova X-rayed some of the moths in question, bred their offspring together, and produced several freak races, among others one with very small wings, which can only crawl instead of flying. Grain infected with the eggs of this race has been sent to Kharkov and seven other centres to breed wingless moths on whose eggs the parasites will live. The parasites can then be let loose into any building where the moth has been seen or its presence is feared. Volkova is now trying to produce a moth without scales, as the scales come off, and make the workers who breed these creatures cough and sneeze.

Now we don't want this sort of thing to happen among our own children, and therefore X-ray workers ought to be protected against doses of X-rays too slight to cause sterility. A sheet of lead is the best protection, and men can be protected fairly easily. Women are a little harder to protect, but it can and should be done where necessary.

Of course all apparatus in the laboratory to which Mr. Willett refers may be so well screened as to make this needless. Unfortunately I am not omniscient and cannot tell him whether this is so. But in view of the number of people who have so far been injured by X-rays, I would advise his daughter to protect herself with a lead apron while at work.

Colliery Explosions

The report on the explosion at Murton colliery, Durham, in June 1942 reminds us of the army which is always in danger of death, both in peace and war, the army of coalminers on whom our industrial life depends. The Murton explosion was not a very big one, but thirteen men were killed.

There is nothing very new to be said about colliery explosions; but as deaths from them are largely preventable, the whole Labour Movement should understand about them, in order to see that the necessary steps are taken as soon as possible.

The Murton explosion, according to the report of the Inspector of Mines, was caused by a multi-shot exploder whose sparks ignited firedamp. Firedamp is the name given to the gas, mostly consisting of methane, which is given off by coal without the application of heat. It is not the same as ordinary lighting gas, which is made by heating coal. It burns with a very pale blue flame, and is hardly poisonous. You can collect it by letting a glass tumbler fill with water, holding it upside down in a pond or ditch with a muddy bottom, poking the mud with a stick, and catching the bubbles as they rise. You can then light the gas; but perhaps you had better wait till there are more matches.

If a stream of pure firedamp is coming out of a pipe in a stopping in a disused pit, one can light it without danger of an explosion. But one must do an analysis first to be sure it is not mixed with air. For mixtures of firedamp and air in certain proportions are explosive. About 9 per cent of firedamp is the minimum explosive amount. In mines where firedamp is a danger, tests are supposed to be made for it constantly. It is fairly easy to test for it with an ordinary safety lamp. One turns the wick till there is only a small blue flame, and the firedamp can be seen burning above it as a pale blue cap. This is quite safe, as the flame is prevented from spreading out and causing an explosion by flameproof screens of wire gauze. Electric lamps give better light than the old oil safety lamps, but unfortunately cannot be used for this test.

Mine inspectors have to learn, among many other things, how to estimate the amount of firedamp in air in this way. Most people can only see the firedamp flame in darkness. My father was very much interested in one candidate who could see it in full daylight. No one else could, but he got the percentage of firedamp right every time, so he was not faking. My father believed that a lot of so-called psychical phenomena could be explained by abnormally acute senses. Thus a famous thought-reader used to go out of a room while the people in it decided on a word or a subject (say a man in a boat) of which they would think. He was certainly out of normal earshot, but he often guessed the thing thought of. As, however, he usually failed when there was a tap running in the house, my father thought that he probably had supernormal hearing, like the inspector's supernormal vision.

The trouble about these tests is that a sudden gush of firedamp may come out of coal, so that the air becomes explosive when it was quite safe half an hour before.

The exploder used at Murton was of a type not approved for use where there is a danger of firedamp. It will be interesting to see if anyone is prosecuted for using it. The firedamp seems only to have caused a small explosion, which by itself might not have killed anyone. But this caused a secondary explosion of coal dust. Air mixed with fine coal dust is very explosive.

Since collieries have been ventilated so that most of the firedamp is carried away, big firedamp explosions have been rare; and most of the large ones have been due to dust.

If enough shale dust is spread along the roads, the mixture of coal and shale dust is not explosive. By the way, the tunnels leading from the shaft to the working face are called roads, even though they are very narrow, and the roof is often too low to let you stand up. But at Murton the dust explosion seems to have occurred in disused roads, where presumably no stone dust had been spread for some time and fine coal dust had drifted in, so that when stirred up and mixed with air it gave an explosive mixture.

The Inspector of Mines recommends that a safer form of exploder should be developed for use in mines where there is any danger of gas. But another lesson can also be drawn from his report. The men were not killed by the force of the explosion, or by burns, but poisoned by carbon monoxide, though some of them may have been so badly burnt that they would have died in any case. Carbon monoxide is always formed in colliery explosions, and sometimes in fires, and is very poisonous. More than forty years ago my father recommended in a report to the Home Office that every miner in mines liable to explode should have an oxygen breathing apparatus similar to those used by rescue squads, but of simpler and cheaper design. This recommendation was never carried out.

An ordinary gas mask will not stop carbon monoxide; but it can be filtered out of air by a substance called hopcalite. Respirators with a hopcalite filter have been successfully used in industries where carbon monoxide is a danger. They would be very much cheaper and lighter than oxygen apparatus, and could probably be developed for the use of miners. This is one of the matters which the Miners' Federation should take up after the war, when the factories now making gas masks will be available for work of this kind. The progress of science is constantly opening new possibilities of detecting poisonous or explosive gases, and of giving protection from them. Trade unions need a first-rate scientific staff to keep in touch with them.

There is still another possibility. Enough firedamp goes up

the upcast shafts of our collieries to light every gas lamp, and probably every gas stove, in Britain. We do not know how to separate it from the air. But in the Soviet Union they liquefy the firedamp which comes up from oil wells in the Baku region, and use it to drive buses. If serious research were done, it is entirely possible that we might be able to use the wasted firedamp from our coalmines, and convert one of the miner's greatest dangers into a valuable commodity. As, however, the Soviet Union is ahead of the rest of the world in methods of separating gases from mixtures, it is likely that the necessary research will be done there if anywhere.

Euthanasia

A correspondent writes to me to ask whether I approve of killing incurable invalids, and whether this is legal in the Soviet Union. She says that her own baby has chronic jaundice, is in constant pain, and that the doctor says it is bound to die. If so, would it not be right to save it unnecessary suffering?

A body called the Euthanasia Society, including a number of well-known doctors, has been formed to urge a change of the law to legalize killing in such cases. On the other hand the Catholic Church, and many non-catholics, hold that it is wrong to kill except as a punishment. To which one might answer that the Church's record of killing for heresy is so black that its opinion carries little weight.

There is no doubt that a vast amount of suffering would be saved if people undergoing great pain from incurable diseases were killed. The argument for killing an idiot child which cannot look after itself, let alone learn to speak, is equally strong. Nevertheless I am against legalizing such killing at present, for three reasons.

The first is the uncertainty of medical diagnosis. Even the best doctors make serious mistakes. They have to guess what is wrong inside us from what we tell them, what they can see, feel, and hear, and sometimes from X-ray photographs or analyses of blood or urine. This is like guessing what is wrong with a motor car engine without lifting the bonnet.

In a medical profession organized as ours is, the ordinary general practitioner does not have the full resources of medical science to help in diagnosis. He has to guess more than is necessary. In fact our medical profession is still largely in the feudal stage of productive relations. We go to a single doctor for a great variety of services, as our ancestors went to the village cobbler for a pair of boots, or the village miller for a sack of flour. The medical profession has not reached the development of productive forces which is made possible by the division of labour reached under capitalism, let alone that possible under socialism. In the hospitals, where there is division of labour, the patients in the wards on the whole get fairly good treatment, but the out-patients are not only treated inefficiently, but often with extreme discourtesy. The future of medicine lies with health centres, run so far as possible by trade unions or other workers' organizations, employing several doctors, and proper equipment for the diagnosis and treatment of disease before it gets to the hospital stage.

So to begin with I would answer that until our medicine is better organized we should not be justified in killing a suffering man, woman, or child because a doctor says they are incurable. My correspondent says that her baby has no gall-bladder, and seems to think this proves it will die. Actually the gall-bladder happens to be one of the organs one can do without; so I wonder if it is quite certain that the baby will not recover.

If euthanasia is ever legalized, I am sure that the killing should not be done by doctors, though some of them seem quite willing to oblige. I cannot think that it would increase most people's confidence in their doctor to know that he had just bumped off one of the neighbours, however kindly he did it.

There is another very big reason against legalizing euthanasia at present. Suppose grannie has an incurable disease and is suffering, her children may have economic motives as well as humanitarian ones for putting her out of the way. Mrs. Smith is living in an overcrowded house, and her death would mean a lot less work for mother, and another bedroom for the children. Mr. Brown has a larger house and a servant, but he will come in for £10,000 under his father's will when grannie dies. It is no good pretending that these economic motives would not weigh with people to some extent.

So I am against legalizing euthanasia until we have a society where there are no overcrowding and no big legacies, a socialist society in which the economic motives which make one man's death into another man's gain are gone for ever.

In *The Origin of the Family*, Engels made a very profound remark about the future of marriage. "That will be answered," he wrote, "when a new generation has grown up, a generation of men who never in their lives have known what it is to buy a woman's surrender with money or any other social instrument of power, a generation of women who have never known what it is to give themselves to a man from any other considerations than real love or to refuse to give themselves to their lover for fear of the economic consequences. When these people are in the world they will care precious little what anybody to-day thinks they ought to do; they will make their own practice and their corresponding public opinion about the practice of each individual, and that will be the end of it."[1]

Just the same is true about euthanasia and many moral questions on which people are divided to-day, for example vegetarianism and the rights of animals. The people of a communist world will differ from us in two ways. They will regard affection as the normal relation between two people, and compassion for the weak as part of human nature. But they will also find one of their main sources of happiness in work for others, and will not want to live when they can no longer do such work.

So these motives will swing them in different ways, and I do not know which will win, though I think some form of suicide may well be legalized. The killing of incurables is not legal in the Soviet Union, though I am sure it would be very leniently dealt with if there were no ulterior motive. And I, for one, am against legalizing it in Britain at present, whatever may be the case in future.

[1] In reply to this article I got several very indignant letters, implying that I was advocating sexual promiscuity. My correspondents apparently thought it inconceivable that, in the absence of legal, religious, or economic pressure, people might actually prefer monogamy. To me, as to Engels, this seems quite probable.

6

INVENTIONS

Inventions that Made Men Free

No one doubts that the great inventions of the last two centuries have revolutionized human society, and profoundly altered the course of history. To that extent everyone is a Marxist. However, opponents of Marxism go on to say that these inventions depended on the development of scientific theory, and that the really revolutionary influence has been that of scientific ideas. There is some truth in this; but only in some kinds of society does theory lead to invention, and it is worth while for Marxists to know something of inventions which were certainly not based on any scientific theories, and which changed the course of history.

After the western part of the Roman Empire collapsed in the fifth century A.D. its territories were occupied by various "barbarian" nations, such as the Angles and Saxons in England, the Franks and Burgundians in France, the east Goths and Lombards in Italy. They were uneducated, but they did not practise an economy based on slavery, and the ordinary man in Europe was probably a good deal freer and nearly as comfortable as his ancestors had been under Rome.

Literature, science, architecture, and so on, were at a very low ebb, but a number of inventions were made which had a great effect on society. Our knowledge of them is largely due to a French cavalry officer called Lefebre de Noëttes, who studied all the pictures and statues of horses, and remnants of harness, dating from more than six hundred years or so ago.

If you look at a picture of a Roman chariot, you find that it was not much bigger than a perambulator, and was pulled by at least four horses. If you look more carefully, you can see why so many were needed. Instead of having harness of a modern type, they pulled it by pressing on a strap in front of their throats. If they had exercised a force of more than a few pounds they would have been throttled.

Further, the Romans did not use iron horse shoes. They used leather ones, or none at all. So the hooves of their horses wore out on paved roads, and would have done so on macadamized or concrete roads. Horses were mainly used in open country. The Romans did not, and could not, use horses for pulling heavy carts. These were dragged by oxen, or by men. The huge stones used in many ancient buildings were largely transported and lifted by human power, often by that of slaves.

Before we start feeling superior to the Romans, we had better remember that in India, South Africa, and other parts of the British Empire and Commonwealth, men are still used as beasts of burden. This may be excusable in mountains or dense forests where horses cannot penetrate. But men pull rickshaws in many towns where there are quite good roads, presumably because they are cheaper than horses.

Some time in the so-called "Dark Ages," very possibly in France, the horse collar and iron horse shoes were invented. This meant that when the Middle Ages began, horses were used for transport in a way which was quite impossible in the Roman Empire. There is one exception which proves the rule. Many of the heavy stones in Chartres cathedral were drawn by teams of men who undertook this hard work deliberately as a penance for their sins. But this was unusual; and the greater use of horses was one reason why slavery was not revived in Europe in the Middle Ages.

Another invention of the Dark Ages was the rudder. The Roman ships were steered either by rowing harder on one side than the other, or by a special pair of oars on each side near the stern, one of which was dipped into the water as required. Even a sailing ship had this pair of oars. This sort of steering may have been all very well in a calm sea with a ship on an even keel. It must have been hopeless when the ship began to roll. Some ships with steering oars were still built at the time of the early crusades, but about this time the rudder superseded them, which made sailing very much easier.

However, it took a long time before wind power completely

ousted man power for moving ships. The Spanish Armada which attacked England in 1588 still included eight galleys and galleasses rowed by slaves, along with one hundred and twenty-four sailing ships; though the English had given up galleys two hundred years or so earlier, for there were none in the fleet in which Henry V sailed to the battle of Agincourt.

The first water mill recorded in Europe was built on the Moselle under the late Roman Empire. The windmill seems to have been another invention of the "Dark Ages," for windmills were being used in several parts of Europe by the twelfth century. The Roman mills had been worked by slaves or donkeys.

Here, then, were a series of most important inventions, all of which served to free men from the most arduous and unskilled work, namely pulling carts and oars, and turning mills. They were made in an unscientific age, and were of great political importance because they abolished completely unskilled occupations in which men were merely sources of power. In fact they were much bigger steps towards human freedom than Magna Carta, the Habeas Corpus Act, or other laws of which we learn at school. Christianity played some part in abolishing slavery, but not a very large one, for both catholics and protestants enslaved negroes. Technology was more important.

One other important invention was made in the "Dark Ages," namely the mechanical clock. This was almost certainly invented in a monastery, while the others were made out of doors. The clock is of great historical importance as the forerunner of every kind of machine in which one wheel transmits power to another, whether through gears, belts, cranks, or worms. In the long run, therefore, it had a very great effect in liberating men from toil. But except in so far as the principle of gearing was used in wind and water mills, its effect as an agent of freedom did not show up for a thousand years.

Technology is to be an important part of education under the new Education Act. If it is properly taught, it can be made the foundation of historical teaching. If it is badly taught, it will be divorced from culture. Every teacher who is even slightly influenced by Marxism should be able to show how human progress has depended on technological improvements.

Polarized Light and Its Uses

We are beginning to think seriously of post-war reconstruction. The efficiency or otherwise of this will depend to a large extent on how far science is utilized. I am quite aware that science cannot be fully applied to human welfare either under competitive or monopoly capitalism. But socialists can at least state how it should be applied.

The loss of life on our roads before the war was appalling, About five hundred people were killed a month, which is more than are killed in air raids at the present time, though, of course, far less than were killed during the blitzes. It is up to us all to see that the death-rate does not rise to the pre-war level when motor vehicles appear on the roads again in their former numbers. Many measures are needed, including better design of roads, crossings, and vehicles, better tests for motorists, shorter hours for lorry drivers, and better education of pedestrians.

I am only going to deal with one, which would save a few hundred lives a year. This is the proper use of polarized light. When a beam of light passes through water or glass it is bent, or refracted, but stays single. When it goes through certain crystals it is not merely bent, but split into two rays, which are polarized. Light consists of electric and magnetic "waves" with a frequency millions of times that of radio waves, but otherwise very like them. Light is polarized when the electric force is always in the same direction. Thus a horizontal beam such as that from a car headlight may be polarized horizontally with the electric force alternately right and left, or vertically, or at some intermediate angle.

One can arrange two bits of crystal in what is called a Nicol prism, so that they will only let through light polarized in a particular direction. If a horizontal beam has come through a Nicol prism transmitting vertically polarized light, it will pass through another set up in the same way. But if we turn the second one through a right angle, no light gets through. We can understand what is happening by analogy. If two men were signalling to one another by sending waves along a taut rope, they could still do so if the rope passed through a vertical slit. The rope would still be able to oscillate up and down. The signals could be passed through two vertical slits. But if the second slit was horizontal, communication would cease.

Within the last ten years it has been possible to make large sheets of material by arranging multitudes of very small crystals between two sheets of glass or celluloid. These screens let through light polarized in one direction only. They are made and sold by the American Polaroid Corporation, and are rapidly replacing Nicol prisms for many purposes.

This is how they can be used to make night driving safer. Headlamps are made so as only to produce horizontally polarized light, and wind screens so as only to let through vertically polar-- ized. A driver sitting behind such a windscreen sees the headlamp of an approaching car or lorry as a faint glow, and is not dazzled. But the light reflected from the road surface and banks is not all horizontally polarized, and much of it gets through the screen. Thus he sees the road lit up by his own or another vehicle's head- lamps, and can avoid collisions. But he is never dazzled, and never has to dip his own headlight. A further advantage is that as the light reflected from a horizontal surface of water on a wet road is mainly polarized horizontally, this source of glare is cut down also.

Clearly this safeguard will only work if all motor vehicles have headlamps and windscreens designed in the same way. This change cannot be made in a moment, and certainly not in war- time. But it could be made compulsory in all new motor vehicles turned out after the war. So long as the old ones remained on the road the full advantage would not be gained. Even polarized head- lamps would have to be dipped. But after a few years the use of polarized lamps and windscreens could be made compulsory.

As our car and lorry factories are now making military vehicles, there will have to be a big switch in production after the war in any case, and this offers an ideal opportunity for such a change. But it will only be possible if the Government sees to it that "polaroid," or whatever material is used, does not become the monopoly of any firm or group in this country, and that excessive prices are not charged for it. Many firms will want to get busy the moment the war is over, and will complain of any State interference. Others may try to get a monopoly. Nevertheless, the change can be made, and thousands of lives saved, provided details are worked out beforehand by those concerned with reconstruction.

Polarized light is already used for many purposes. When a polarized beam passes through a sugar solution the plane of polarization turns through an angle depending on the amount of sugar. This is the most accurate method of measuring the amount of sugar present, and is used by analysts for sugar and many other substances.

Again, polarized light can be used for secret signalling. If the observer looks through a Nicol prism, and the transmitter has a prism on his signalling lamp which he turns through a right angle, the light will go on and off, although an observer without a prism sees no change in its intensity. In 1939 many people thought the towns would be better hidden if a few lights were allowed in country districts. One reason why this was not allowed was probably the danger that fascists in rural districts would use polarized light to signal to enemy planes.

I have devoted an article to a very minor point in reconstruction as an example of the need to apply recent scientific and technical developments on the largest possible scale. This would of course be vastly easier under socialism than capitalism. The Soviet Union and other European States which adopt socialism after the war will find reconstruction far simpler than the capitalist States. I am willing to bet that unless the election which we are promised after the war gives us a clear majority of real socialists, this particular application of science will not be made in Britain until other countries have given the example. Private interests of one sort or another will block it, and a few hundred lives will be lost each year in consequence.

There are, of course, vastly bigger and better reasons than this, for desiring a socialist Britain as soon as possible. But the possibility of rapid technical improvements is an argument for socialism, and not a negligible one.

The Spectroscope

Among the instruments which are now again at my disposal since I have got back to University College, London, is my pocket spectroscope. It consists of a set of prisms which bend the light coming through a slit at one end through angles which are different for the different colours, and a lens to focus them. I carry it in my pocket in the hope of seeing the spectrum of a flash from the explosion of a V1. Unfortunately the V2 gives no warning, so I have little chance of analysing the light from it. Meanwhile I amuse myself by looking at other lights, and probably bore my colleagues by asking them to do so.

An ordinary electric lamp shows the various colours of the rainbow, from red through orange, yellow, green, and blue, to violet; though the violet is much fainter than in sunlight. But in peace-time one can see a great many lights of a quite different character in the streets of any large town. If you look at the yellow sodium lights which are used in some streets the spectroscope does not show a band of all colours, but two yellow lines close together. At present the easiest spectrum of this kind to see is that of neon. The tube trains, besides their ordinary lamps, have little lamps containing this gas, and giving out a pink glow. If you look at a neon lamp through a spectroscope you see a number of red lines, a yellow one, and two rather dim green ones. In peace-time one can also see the characteristic bright lines of mercury, nitrogen, and other elements used for advertisements, and outside cinemas.

A hot solid or an ordinary flame gives out light of all the visible wave-lengths. A gas such as neon, or the vapour of sodium or mercury, gives out light of a few colours only, when excited electrically or when heated very strongly. If one looks at a strong electric lamp through the glow of a weak neon lamp, the same lines which appeared bright before, now appear dark. That is to say neon atoms take up from strong light the same kinds of rays which they give out into darkness.

The spectroscope has been of great practical value. Every element, when brought into the state of hot gas, gives out its own set of colours, showing as a bright line spectrum. Its quantity can often be estimated with fair accuracy from the intensity of the light, and so a photograph of the spectrum, say of an alloy or the salts extracted from soil, enables us to state its composition.

The spectroscope is particularly useful for detecting rather small amounts of an element, and has shown that various plants and animals collect certain elements with extraordinary efficiency. For example, Professor Fox, now at Bedford College, London, and his colleague Ramage, showed that the ash of ordinary mushrooms contains substantial quantities of silver, though not enough to make it worth while ashing them as a source of this metal. The common scallop collects cadmium, a metal used in paint manufacture, from sea water; and a sea squirt collects the rarer metal vanadium, which is used for toughening steel. Until these animals were analysed it was not even known that salts of these metals existed in the sea. Vernadsky and his pupils in the Soviet Union think that some mineral deposits containing rare elements were formed by animals and plants which concentrated them in this way.

The spectroscope has also given us a great deal of information about the sun and stars. The pioneer in this research was Lockyer, who discovered that most of the elements known on earth were also found in the sun. And one, the rare gas helium, was detected in the sun before it was found on the earth. The sun shows dark lines because the gases in its atmosphere stop certain of the rays coming from its glowing interior, like the neon lamp. During an eclipse there is a moment when the sun's interior is covered by the moon, but its atmosphere is visible. The light from the gases in it then appears as a series of bright lines.

The Indian physicist Saha pushed the analysis further, and showed how the relative intensities of different spectral lines can be used to measure the temperature and pressure of the atmospheres round stars. In consequence we know a good deal about the physical state of stars, as well as their chemical composition. But the most important service of spectroscopy to science has been to show how atoms and molecules are constituted. Why does a hot atom of neon give out light of only a few colours, instead of the whole possible range of colours, like a white-hot electric bulb filament? Einstein showed that light of a particular colour is given out or absorbed in packets called photons, the amount of energy in a photon being proportional to the number of vibrations of the light per second. And a particular atom gives out its energy in characteristic units.

A man can only pay a bill in units which are different in different countries. The smallest English unit is a farthing, the smallest American a cent, and so on. So if you know that a man always pays his bill in cents, you know that he lives in America, and so on. A large-sized material object, such as a fly-wheel, can have any energy over a wide range, and can give it out in a continuous stream. But an atom or a molecule is like a fly-wheel which can only rotate at certain definite speeds, and therefore if it gives out energy or absorbs it, must do so in standardized quantities. Each spectral colour represents the difference between two of these energy levels.

When this principle was understood it became possible to make tables of the energy levels of an atom or molecule. Some levels are due to vibrations or rotations of a whole molecule or a part of it, others to electrons in it moving at one or other of a standard set of speeds. Once the spectrum of an atom or molecule is catalogued and the energy levels calculated it becomes possible to predict its chemical behaviour. It is of course necessary to catalogue the spectrum in the invisible but photographable ultraviolet and infra-red regions, as well as the visible one.

From the spectrum of hydrogen one can calculate that two hydrogen atoms will unite to form a molecule, and how much energy will be given out in the process; while the spectra of mercury and chlorine explain why two mercury atoms will not unite with one another, though either will unite with chlorine, and so on. Thus chemistry is becoming a rational science, not a mere collection of rules. Unfortunately the mathematics involved in the calculations is still so terribly complicated, that it is less trouble to learn the rules than the general laws behind them. However, the calculations have been justified and shown to be this-sided by predicting a number of previously unknown chemical facts.

The positivistic world view is that science reveals a number of laws connecting our sensations, and that it is no use to try and go deeper, and explain these laws. Thus Comte thought that chemistry could never be reduced to physics. The view of dialectical materialism is that nothing is isolated or inexplicable, and that therefore chemistry and physics can be unified; though, as Lenin wrote, the result will be that apparently simple things, like electrons, will turn out to be much more complicated than they first seemed to be. There is no doubt that this second view is not only much more stimulating to research workers, but is being more and more confirmed by discovery. And the spectroscope has played a major part in this unification of science.

The Electron Microscope

In 1898 J. J. Thomson published the results of some years of work on the passage of electricity through a vacuum. By bending the current out of its path both with an electric and with a magnetic field, he showed that it consisted of negatively charged particles which were later called electrons. This discovery was of immense importance both for the development of physics and chemistry, and in practice. To take only one example, a radio valve depends on the fact that a stream of electrons shot out of a hot wire can be regulated by an electric field.

One of the most recent applications of streams of electrons is to microscopy. A microscope using light will not magnify things clearly more than about 3,000 times, whilst an electron microscope can already magnify up to 300,000 times, and will probably do much better in future. So the electron microscope may lead to almost as great advances in knowledge as did the ordinary microscope. It works in much the same way as an ordinary microscope. A magnet attracts a beam of electrons as it attracts a wire carrying a current in a dynamo. A ring-shaped magnet suitably designed will focus a parallel beam to a point, just as a lens focusses a beam of parallel light rays. Of course the magnet needs as careful design as a lens.

A modern electron microscope is about seven feet high, and

the cheapest costs over £3,000. At the top is a cathode firing off electrons at 60,000 volts. The beam passes through three magnetic "lenses," corresponding to those of the condenser, the objective, and the ocular in an ordinary microscope, and through the object to be photographed. The whole action takes place in a very high vacuum, for air scatters electrons as fog scatters light. Special controls keep the electron voltage and the current through the objective lens steady to one part in 50,000. The specimen is mounted, not on a glass slide, but on a very thin film of cellulose.

The technique has been used in industrial chemistry. Apply the beam to the smallest dot on a photographic film, which looks like a faint cloud under an ordinary microscope, and we see a complicated tangle like a snarl of string. This result is being used for the improvement of very fine plates such as are needed for astronomy and aerial photography. Other industrial chemists are studying thin films of metal, and others again the synthetic fibres such as nylon and vinyon which are replacing silk and rayon, and the plastics which are being used in so many branches of industry.

This microscope has definitely proved that the molecular theory of chemistry is true, for large protein molecules have been photographed, and their size and shape found to be the same as had been determined by other less direct methods.

Probably its most important applications will be in biology. Bacteria which are so small that the ordinary microscope cannot distinguish their parts, turn out to have quite a complicated structure. Viruses, agents of disease far too small to be visible with the light microscope, have been photographed, and different kinds can be distinguished. So far as I know, no photographs have yet been taken to show the detailed structure of human cells, but this will doubtless be done soon if it has not been done already.

The importance of this new tool for research lies in the fact that we still know very little about the structure of things smaller than the smallest cells we can see which are alive, but larger than the largest chemical compounds we can make which are not alive. Certainly some of the larger molecules show a few properties which are generally found in living things, and the smallest organisms do not possess all the qualities which we generally associate with life. But there is a considerable gap, and there are still people who say it is unbridgeable.

The history of science, and especially our knowledge of evolution, should make us very cautious about saying that any distinctions such as that between living and dead matter are absolute. The electron microscope is already helping us to bridge this particular gap.

The Reading Machine

An invention has recently been made which may have as big an effect on the spreading of knowledge, and for that matter of lies, as did the invention of printing. We are apt to think of books as part of the natural order of things, though the oldest printed books are not five hundred years old. It is quite likely that some of my readers will live till books are curiosities only consulted by antiquaries.

The history of books is a remarkable illustration of human conservatism. The Sumerians, Babylonians, and other peoples of ancient Iraq wrote on clay tablets which they baked. These are extremely durable, and vast numbers have been dug up. But they are also very bulky. A learned king had to use hundreds of slaves to make a special wing for his palace to house what one of us can get on an ordinary bookshelf. The ancient Egyptians made a great improvement by using papyrus. This is made from strips of a rush gummed together, and is the origin of our word "paper." Unfortunately they did not think of binding pages together, but made it into continuous rolls. As one read through a book one unrolled it from one peg and rolled it on another like a cinema film. The Romans largely replaced papyrus by parchment, that is to say fine leather, and books of the modern type were first made in Europe about A.D. 100. The Chinese have certainly used paper for 2,000 years. The Arabs learned to make paper from the Chinese about A.D. 750, and the Europeans from the Arabs about A.D. 1200. It was made in factories from the first, and was one of the few commodities always made by mass production even in the middle ages.

Printing, like paper, was invented in China; the oldest printed book is dated A.D. 868. In A.D. 1041 Pi Sheng invented movable types, but as the Chinese language needed several thousand different characters, they were not much used. In Europe movable type was invented soon after the introduction of printing in the fifteenth century. It is amazing that seals were known in Iraq as early as 2000 B.C. and were used on clay tablets; but no one seems to have thought of applying the same principle to paper for thousands of years. Still more remarkably quite early seals were cylindrical, and were rolled on the clay tablet. But the rotary press was only invented in A.D. 1790, and first used, for printing *The Times*, in 1814.

The new invention is this. An entire book is photographed on a film. This may be an ordinary photographic or cinema film, or a special micro-film. In either case it is about an inch across. It is quite thin, and far too small to read directly. So its image is projected onto a screen with an electric light. The reading machine is about two feet high, and can be stood on an ordinary table. At present it costs about £15 and is not on sale, though a few have been given by the Rockefeller Foundation to British libraries. The revolutionary fact is the extreme smallness of the films. A whole book rolls up into a case a good deal smaller than a reel of cotton. You could carry the *Encyclopedia Britannica* in one pocket, and the whole library of the British Museum could be stored in a fair-sized house. It is curious that after nearly two thousand years we should have gone back to rolls, though on a much smaller scale.

Micro-films have been used for some years in America, particularly for scientific publications. But in spite of the efforts of Mr. Watson Davis, of the American Science Service, most people regarded them as an amusing toy rather than a serious invention.

But the war has altered this. It is impossible to get European scientific journals in any numbers, though single sets of many can be got through Portugal, Turkey, or Sweden. But they can be photographed on micro-films. Reading machines are now available in the Science Library in London, among other places; and these journals can be read from micro-films, of which there are a number of copies. Once the micro-film habit has caught on, some American scientists hope to publish journals on micro-film only. It is claimed that this will be cheaper than printing. It will certainly be cheaper for the libraries which have to store them. On the other hand it is doubtful whether films of the present type will last as long as paper.

As soon as paper is given up for micro-film there will be considerable changes in the arrangement of books, and probably in literary style. However, the social effects will be still more important. It will be entirely possible for a small town library to have 100,000 films, each representing an entire book or volume of a periodical. And reading machines will be a good deal cheaper than radio sets. This will mean far greater opportunities of culture for the masses.

But these opportunities may not be given. The ruling class may try to monopolize the reading machine as it has monopolized the radio, and very nearly monopolized the press. Unless the Labour Movement wakes up to the situation before the capitalists it may be as difficult to get micro-films of Lenin's works as to hear a communist speaker on the radio. Or the reading machine may be effectively kept for the well-to-do, while the general public is spoon-fed with books guaranteed to raise no dangerous thoughts. Every new invention is a chance for the workers and a chance for the bosses. The price of liberty is eternal vigilance.

Listening to Doodlebugs

If Londoners learned nothing else between June and September 1944 they learned that light travels faster than sound. The most striking demonstration of this fact was to stand on Parliament Hill and watch the bombs bursting in London. One saw a doodlebug burst in Wandsworth, and heard the burst half a minute later.

Light travels at the enormous speed of nearly a hundred and ninety thousand miles per second. If you had a system of mirrors going round the world, a flash of light would take under a seventh of a second to go round them and return to its starting point. The time taken for light to cross London can be neglected for all practical purposes. This does not mean that we see a thing as soon as it happens. When the light strikes the back of our eye it breaks down a purple substance called rhodopsin. This starts messages along a number of nerve fibres, and these are switched over to other fibres, finally reaching the area at the back of the brain concerned in vision. This takes about a tenth of a second.

Sound travels at the moderate speed of 750 miles per hour, or roughly a mile in five seconds. So with a stop-watch one could quite easily estimate the distance of a doodlebug burst within 400 yards. The doodlebug travels at about half the speed of sound. So its sound precedes it and gives a warning. A howitzer or mortar shell also travels slower than sound. On the other hand a field gun shell or bullet from a rifle or machine gun, let alone an anti-aircraft shell, travels quicker than sound, and gives no warning. A rocket may travel either slower than sound, or faster. V2 goes a good deal faster, and therefore gives no warning. A hooter on a car travelling as fast as sound, or faster, would be useless.

Sound consists of series of pressure waves moving through the air. The air consists of rapidly moving molecules. They are travelling at many different speeds. Some are moving in the same direction as the sound; others in the opposite direction, sideways or obliquely. So the speed of sound is a good deal less than the average speed of the molecules, in fact about 74 per cent of it. In a gas the molecules are far apart, but in a liquid there is not much space between them. So the sound travels much quicker. Roughly speaking, the time taken is equal to the length of the gaps between molecules, divided by their average forward speed.

In fact sound travels four and a quarter times as fast in water as in air. This has an important bearing on the hunting of submarines by the "Asdic" method which depends on sound. Sound is not much good for locating an aeroplane. The sound from a plane three miles away takes fifteen seconds to reach our ears. During this time a plane moving at 300 m.p.h. has gone a mile and a quarter. But a submarine under water moves at about 10 or 15 m.p.h., and sound travels faster in water. So sound location is about one hundred times as efficient against submarines as against aeroplanes. It is in fact very useful, even though a corvette or frigate chasing a submarine aims a small distance behind it.

Except in dealing with distant objects producing a great noise, such as aeroplanes, the lag in transmission of light or sound makes very little practical difference to our actions. The lag between excitation of a sense organ and muscular action is much more serious. (One cannot determine the lag between sensation and action exactly, because there is no way of measuring, within a split second, when a sensation begins).

Simple reactions to simple stimuli do not take very long. If a man is told to press the button when he sees the light, the reaction time is about a tenth to a fifth of a second. If he is told to press the button when he sees a red light and the pedal when he sees a green one, about two-fifths of a second are needed; and a good deal longer for more complicated reactions, though these times can be reduced by training. A fighter may move over 20 yards in a tenth of a second. So not only must a fighter's pilot's reaction times be as short as possible, but there must be no lag in transmitting them to the rudders and wing tips.

Societies react very slowly to new situations, and religious bodies are even slower than political ones. In particular states react very slowly to changes in productive forces. In England we still have large vestiges of feudalism, such as the House of Lords and hereditary ownership of land. Feudalism worked well enough when every manor produced its own food and clothes. It was already out of date when traders could use pack horses or ox-carts, but it survives into the age of aeroplanes and railways. Over most of the world capitalism survives. This again was quite efficient in the early stages of the development of trade and manufacture, but was already out of date a century ago. Today it will only work at all if its "normal" working is interfered with by an elaborate system of controls.

Socialists are aware that capitalism is out of date, and most socialists desire to sweep away many other out-of-date institutions. But they do not always realize the full possibilities of technical progress. For example, pre-fabricated houses far larger and more durable than the Portal house have been made in small numbers, and should form a part of our housing programme. Our methods of heating houses could be overhauled, with a great saving of coal and gain in cleanliness. Our cleaning methods, both in the house and the scullery, are still in the feudal stage.

A socialist should make himself aware of the improvements which technical progress has made possible, not only in society as a whole, but in the details of life. If he does not he is in the position of an anti-aircraft gunner who, instead of aiming ahead of a bomber, aims in the direction from which its sound comes.

Farming the Sea

The fertility of land depends on the presence of sufficient nitrogen, phosphorus, potassium, and other elements, in the soil. Many primitive methods of agriculture exhaust one or more of these elements, and the farmers then abandon their fields and move elsewhere. A society based on this kind of agriculture is not much more developed than one based on hunting. Our ancestors overcame soil exhaustion to some extent by rotation of crops, by the use of manure, and so on. We have greatly increased productivity by using fertilizers such as superphosphate and sulphate of ammonia, and lime not only to supply calcium but to overcome acidity.

An acre of water may produce as much food as an acre of land, or more. But it is even more easily exhausted of essential mineral constituents. In Europe carp are frequently kept in ponds and fed with hay which is dumped into them, but there is very little fish-farming of this kind in England.

The sea yields far less fish per acre than a very poor pasture will yield in meat or milk. What is more, the yield of fish is already as high as possible in many areas, and more intensive fishing may actually diminish the catch. This is because the sea is very short of two essential elements, nitrogen and phosphorus. Its yield of food per acre never rises to the level reached in a lake where these elements are abundant. And even if the phosphates were available, it would be impossible to keep up a high phosphorus level in sea water, as it is precipitated as calcium phosphate.

However, an experiment now under way, and on which a preliminary report has been made, proves that fertilizers can be used effectively in sea water. The site chosen was Loch Craiglin, a lake of 18 acres in Argyllshire, communicating with the sea by a narrow channel at high tide, and somewhat brackish in its upper layers.

The crop consisted of flatfish, namely plaice and flounders. During the year beginning in April 1942, 600 lb. of sodium nitrate and 400 lb. of superphosphate were thrown into the lake, and during the next year rather more nitrate and the same amount of phosphate were added. The immediate effect of a fertilization was that within three days the numbers of small green single-celled swimming plants had more than doubled. These plants are the ultimate source of food for almost all the animals in the sea. Seaweed and other plants big enough to be seen without a microscope are almost confined to the shores, and are quite unimportant as a source of food. The microscopic plants are eaten by small crustacea, the swimming larvae of molluscs and worms, and filter feeders such as oysters, which strain water through their gills. The smallest of them are smaller than blood corpuscles, and their numbers in sea water may rise to thirty million per cubic inch of sea water, and reached four times that number in the fertilized lake. The animals which eat them were eaten in their turn, and there was a great increase in the animals living on the bottom, such as cockles, worms, small crustaceans, and midge larvae.

Twenty-five thousand flounders and six hundred plaice were transferred into Loch Craiglin, which originally contained hardly any. A number were weighed and measured, and all fish caught later were also measured. It was found that the growth rate was enormously greater in the fertilized lake than in the sea. In fact the young flounders put on weight sixteen times as fast as those in the sea loch from which they had been taken. The plaice also did very well. In particular, both plaice and flounders went on growing during the winter, which they do not normally do, apparently owing to shortage of food.

The investigation was not without difficulties. Eels ate some of the flatfish, and a flock of cormorants decided they were on to a good thing, and took their share. Finally the local fishermen, who had originally laughed at the scheme, changed their opinion, and it is thought that some fish went into their creels.

The idea of this work was mainly due to Dr. Gross, a German refugee who is an expert on growing small sea animals in laboratories, and thought his work could be applied on a big scale. He was encouraged by Professor Ritchie of Edinburgh University, and partly financed by Imperial Chemical Industries, who are naturally looking for new markets. The work, which included counts of animals and plants of all sizes from a ten-thousandth of an inch to four feet long, was carried out by Gross, Raymont, Marshall, Orr, Nutman, and Gould working as a team, largely in the intervals of other research. They believe that it would be well worth while to fertilize the water in arms of the sea such as the lochs of Argyllshire, where the fertilizer would be so rapidly converted into plants and animals that most of it would not drift out to sea. They look to a future, to quote their words in an article in *Nature*, "when fisheries will follow the path of agriculture, when development and production will take the place of conservation and restriction."

At present the sewage of our great towns is so treated that much of the nitrogen is lost, while the phosphorus is in an insoluble form. With a different treatment this need not be the case; and the sewage of Glasgow, for example, after proper treatment, might be conveyed by pipes to fertilize the waters of the Western Scottish lochs.

Perhaps Dr. Gross and his colleagues have founded a new industry. It will only give its best yield if fisheries are rationalized both on a national and an international level. The sea is not private property, and the State will have to pay for fertilizing it. We must see to it that this results in cheaper fish and better earnings for fishermen, not bigger profits for middlemen. And the peace settlement should include an international control of fisheries. It is little use trying to use the Moray Firth as a breeding place for fish if foreign fishermen can trawl there to any extent; and the Soviet Government has similar objections to foreign fishing in the White Sea.

It seems likely that fish-farming can be a big new source of cheap and good food. But this will only be so if cut-throat competition is avoided, and scientific planning goes with scientific fertilization.

A Substitute for Morphine?

Research on subjects unconnected with the war has slowed down greatly in Britain, as everywhere else except perhaps in Sweden and Switzerland. But some very interesting work is still being published. A particularly hopeful discovery has just been made by Professor Dodds of the Middlesex Hospital, London, and his colleagues Lawson and Williams.

Morphine is one of the most valuable of all drugs. It abolishes pain completely, or reduces it to a level where it is no longer distressing. By doing so it has saved thousands of lives, by allowing people to get rest and sleep which would otherwise have been impossible.

Some substances easily made from it are also valuable. For example, diacetyl-morphine, or heroin, is not only a pain-killer, but a specific against coughing. But morphine and its derivatives have grave disadvantages too. Besides depressing the activity of the part of the brain (probably the thalamus near its centre) concerned in producing pain, they upset that of the cerebral cortex, concerned in accurate perception, willing, and thought. And they damp down important reflex actions such as breathing. One dare not give morphine to a patient in great pain if there is any danger that his breathing may stop.

Worst of all, they are drugs of addiction. If a patient is given morphine or heroin for some weeks the dose needed to deaden pain rises steadily, until he can take a quantity which would kill a normal man or woman. And when the dose is stopped many people become extremely miserable, even though the pain is gone, and will lie, steal, and even murder to get more. It is worth adding that not everyone would become an addict. I don't like the way morphine upsets the working of my mind. If it must be upset, I much prefer the effects of beer or whisky. And after a fortnight or so on regular doses of heroin I had no discomfort on stopping it. I expect most people would resemble me. Still there are enough potential addicts to make a very rigid control of the sales of these drugs necessary.

The constitution of morphine, that is to say the way in which its seventeen carbon, one nitrogen, three oxygen, and nineteen hydrogen atoms are arranged, has been known for a good many years. But it has not yet been possible to make it from simpler substances, partly because the arrangement of the atoms is a rather complicated one, which cannot be adequately represented on paper, but needs a solid model.

When a natural drug can be made in the laboratory, one can also make others of slightly different pattern, and hope that some of them will have properties of the same kind, but rather more useful. Thus the natural drug ephedrine raises the blood pressure and keeps one awake. But the related synthetic benzedrine has far more effect on the brain for a given effect on the blood pressure.

Dodds has started a new attack on the problem of synthetic drugs. Forty years ago some chemists thought, with Ostwald, that chemical formulae were mere short-hand, and not rough pictures of molecules. The majority probably agreed with Lenin that they were true if rough pictures. The Braggs showed by means of X-ray photographs that this was so. But there is one important exception. In most chemical textbooks open chains of carbon atoms, as in the paraffins, are represented as straight. Actually they can be stretched fairly straight, but are zigzag or curly when relaxed. So a structure built partly of open chains of carbon atoms may have the same shape as one built wholly of rings.

Dodds introduced this principle into pharmacology by making stilboestrol, a compound whose molecule has much the same shape and size as that of the female sex hormone, oestrone, though its formula is decidedly different. Much to most biologists' surprise, it proved rather more effective than the natural product, and is used instead of it in agriculture and medicine.

Now he has done the same with morphine. He tried out sixteen compounds all much simpler than morphine, but with molecules of much the same shape. All were less effective per unit weight. That is to say it took a lot more to kill a rat, or to make it a little unsteady on its legs, which is the effect of a dose which does not endanger life. Some of these compounds were then tried on human beings. Tests were first made on healthy people to see what doses could safely be given. Some were turned down because they caused a good deal of mental confusion. In order to test their properties as pain killers they were tried on patients with inoperable cancer, who were already being given morphine, and suffered severely if it was withheld for even four hours. One of the compounds has so far given complete relief without any mental confusion.

The most curious thing about it is perhaps that it was made in 1887, and in the intervening fifty-seven years no one had apparently suspected that it might be a valuable drug. There was indeed no reason to suspect it until the real shapes of molecules were discovered.

A lot more work will have to be done before one can say that Dodds and his colleagues have got hold of a drug as useful as morphine, let alone a better one. But since it will be possible to make a great many compounds fairly closely related to β-hydroxy-βa-diphenylethylamine, their best pain killer, it is highly probable that within a few years we shall have a drug which, for a given efficiency in stopping pain, will have less effect than morphine in upsetting thought.

It is also possible that any competent chemist may be in a position to make half a ton of a habit-forming drug as bad as morphine, or worse, from easily obtained raw materials, without even breaking the law. Every advance of science opens up new possibilities of good or evil. In this case the evil ones are quite likely to predominate for some years unless action is taken fairly rapidly to prevent the sale of these compounds in the sacred name of private enterprise, until they have been conclusively shown not to produce drug addiction.

NOTE.—Since this was written, Professor Dodds has found that β-hydroxy-$a\beta$-diphenylethylamine is nothing like as good an all-round suppressor of pain as morphine, but is particularly valuable against the terrible pain which may be caused by pressure on nerve trunks, which is unfortunately a common result of cancer.

7

SOVIET SCIENCE AND NAZI SCIENCE

The Cruise of the "Sedov"

IN 1938 three Soviet ice-breaking steamers, the *Sadko, Malygin,* and *Sedov,* were caught in the ice to the north of Eastern Siberia. The larger ice-breaker *Yermak* reached them, and the *Sadko* and *Malygin* were able to get away. But the steering gear of the *Sedov* was damaged, so she was left in the ice, and the crew were told not only to preserve their ship but to make scientific observations. The ship drifted till January 1940 when she was rescued by the ice-breaker *J. Stalin.* The first scientific results of the voyage were read to the U.S.S.R. Academy of Sciences in April 1940 by Buynitsky and Efremov.

The crew had to face great difficulties. The work of sounding had been done on the *Sadko* with an electric winch, and there was no time to trans-ship it. The *Sedov's* crew had to make a winch for themselves, and to join various lengths of cable together till they got a cable $3\frac{3}{4}$ miles long. "It took our mechanics some time," writes Buynitsky, "before they found a method of joining the ends of the wires so that the cable should remain strong and flexible at the joints. All this work had to be done in the dreary cold with no other light than the dull glimmer of a lantern." It looks as the *Sedov's* crew had gone some way towards realizing Lenin's ideal of "men who can do everything."

The first task was, of course, to ascertain the ship's position by observing the sun, moon, or stars. She drifted so far north as to go within about two hundred miles of the north pole, so that on the whole her course lay between that of Nansen's ship, the *Fram,* begun in 1894, and that of Papanin's camp which was landed from aeroplanes at the pole in 1937. Both the recent expeditions drifted faster than Nansen's ship. This means that the whole polar icefield is probably moving faster, and in particular, emptying faster into the Atlantic. This is probably due to a change in the average speed or direction of the wind. But such changes may, in the long run, have big effects on climate.

The soundings show that the bed of the Arctic Ocean is more irregular than was thought, and a record depth of over 17,000 feet was measured. Such observations will ultimately be needed to understand, among other things, the currents at different depths in this ocean. Samples of the sea bed were taken at the same time as the depths were measured.

To find out what sort of rocks there are below the ocean, measurements of the force of gravity were made with pendulums. The stronger the earth's attractive force, the quicker a pendulum swings. In fact a pendulum clock which is right in London loses about $2 \cdot 3$ minutes per day at the equator, and gains about $1 \cdot 5$ minutes near the pole. For at the equator some of the earth's pull is balanced by the centrifugal force due to its rotation. In unfrozen seas, pendulum observations have only been made in submarines, since ordinary ships roll and pitch too much. In fact this is the only constructive purpose which submarines have served. But observations can also be made on a ship stuck in the ice. The observations made on the *Sedov* have not yet been worked out, but they will probably show the same as the submarine observations. That is to say, the force of gravity will prove to be much the same as on land. This is believed to be because the rocks under the sea are denser than those of the continents, which makes up for the fact that water is lighter than rock and exerts less pull on the pendulum. In fact the continents stick up because they are made of lighter material floating on the same semi-liquid material as the heavier rocks under the oceans. Of course it is not yet sure that the Arctic Ocean conforms to the general rule.

In order to measure the earth's magnetism ice houses were set up a quarter of a mile away from the ship, to avoid the effect of the metal in the vessel. The compass was found to be so unsteady that on one day it deviated by fifty-two degrees from its usual direction, and even on "calm" days it was often two degrees out. One reason for this extreme variation was the intensity of the aurora borealis, or northern lights, which were also observed regularly. When the sky was clear during the long winter night they were so bright as to cast shadows. As the northern lights are caused by an electrical discharge, they naturally affect the compass, and the relation between them and the vagaries of the compass was examined. The voyage took place at a time when auroral activity was near its maximum. In ordinary years the compass would be steadier.

The usual weather observations were made, including temperature, barometric pressure, wind direction and speed, and rainfall and snowfall. These were telegraphed to the Bureau of the Northern Sea Route Administration in Moscow. If these reports prove of value to navigators, it is possible that in future several ships or floating camps may be constantly kept in the Arctic Ocean at any given time.

Although the crew included no professional hydrologist, Efremov took the temperature of the water at different levels at forty-three points. Samples of water were also taken and sealed up in every available glass jar and bottle on the ship. These are now being analysed. The temperature measurements show that the water, down to a depth which may be over six hundred feet, is below freezing-point. I mean, of course, the freezing point of fresh water, for salt water does not freeze till a lower temperature. Then comes a layer of comparatively warm water which drifts in from the Atlantic. Below 2,400 feet the water is again below freezing point.

These discoveries were made under very great difficulties. The observations at depths down to 2,300 feet were made with a hand windlass, and a crew of three assistants took turns in working it, which doubtless kept them warm even during the polar night.

Some zoological observations were made. Seals, narwhals, and gulls were seen during the summer of 1938, but in that of 1939 only a few polar bears, and starving and exhausted birds which had been driven north by gales.

Such work as this completely refutes the statement sometimes made that Soviet science is only concerned with finding out facts which will be immediately useful. On the contrary these discoveries will ultimately fit into a general knowledge of our planet which will some day benefit all men. Two hundred years ago British explorers took the lead in exploration from the Spanish and Portuguese. Men like Cook, Livingstone, and Scott made Britain illustrious. Today the torch is in the hands of the young socialist workers of the U.S.S.R.

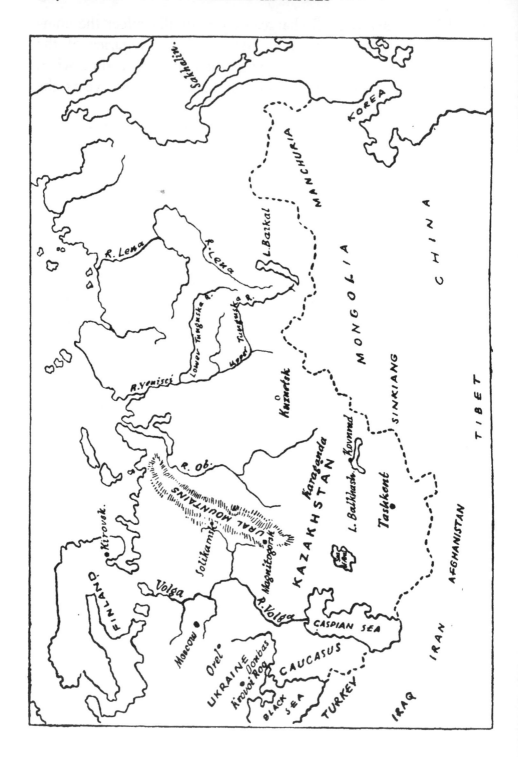

How Two Thousand Geologists
Saved the World

In their preparation for the assault which took Orel the Red Army is reported to have fired ten times the weight of shells per hour that the French fired at Verdun in 1918. And yet the main centre of Soviet steel production before the war was in the Ukraine, using the coal of the Donetz basin and the iron ore of Krivoi Rog. The Nazis conquered this area in 1941, and still hold it.[1]

How has this extraordinary recovery been possible? There are three reasons. One is the extreme efficiency of the Soviet economic system, the system of socialism which, according to anti-socialists, "cannot possibly work." The second is the devotion of the Soviet mine, factory, and transport workers, who are toiling under terrible conditions to preserve their way of life.

But these would not have been enough without a third factor, namely Soviet science. Most of the details of how science is applied in the Soviet factories are secret, as they are in Britain. But we do know something of the work of the geologists who made it possible to move industry east. If the Midlands of England were invaded, it would be no use moving steelworks to the Scottish Highlands, for there is no coal there, and not much iron. It would be little use moving them to Lancashire, Durham, or the Clyde, for the output of coal could not be doubled in a year, even if all the miners were moved also.

Lenin saw that his country could not be industrialized without a full knowledge of its geology. The Soviet Government rapidly built up the world's greatest geological survey. By 1940 it employed two thousand fully trained geologists, besides about eight thousand assistants, and cost a thousand million roubles a year. If you take the rouble's purchasing power at about sixpence, this is £25,000,000; and I think this is a fair estimate, based on the price of bread, though before the war the rouble went much further than sixpence in buying books or theatre tickets, and not so far with clothes.

---

[1] It was reconquered in 1943.

The British geological survey, if I remember, employed ten geologists in the field; though we still have a lot to find out, for example, the extent of the Kent coalfield. The geology of the British Empire is largely unknown, and much of the knowledge is a secret of mining and oil companies. No area of comparable size has been half as well studied as the Soviet Union.

Apart from the Lena goldfields, the only minerals in Siberia which were being mined on any great scale before the Revolution were the coals of the Kuznetz basin, but even this had been barely scratched. It is now believed that this area and the Tungus basin contain enough coal to last the whole world at its present rate of consumption for seven centuries. The Karaganda coalfield is smaller, but very important because it lies between two great metalliferous areas. The Tungus and Karaganda coalfields were only found after the Revolution. The Karaganda coal is used to smelt the copper of Kounrad, also discovered by Soviet geologists, and the iron and other metals of the Ural region round Magnitogorsk, which was enormously developed after the Revolution.

The Caucasian oilfields were, of course, known before the Revolution, but the new oilfield discovered between the Urals and the Volga was of great importance when the Nazi thrust to Stalingrad cut the main lines of supply from the Caucasus.

Supplies of many other metals were discovered; for example, magnesium, which is used for aeroplane construction and for firebombs, at Solikamsk, along with the world's biggest potash deposit. Even tin, which is one of the few metals of which the Union is short, has been found in Siberia.

"Move industry east" was one of the tasks of the second and third Five Year Plans, not merely as a safeguard against invasion but to equalize the development of different nations in the Soviet Union. It was not merely uneconomic to take steel for thousands of miles from Ukraine to Siberia, when it could be made on the spot, or to take raw cotton from Tashkent to Moscow, and send cotton goods back again. Stalin saw that the formerly subject peoples had a right to industrial development. The war speeded up this movement vastly. Factory equipment and workers from Ukraine were loaded into trains and put down in Kazakhstan or

Siberia to start new factories. This was possible because the newly discovered minerals were near the surface. A hundred and fifty years ago we were still working coal outcrops in Britain. The coal was a few feet below the surface, and no deep shafts had to be sunk. This is the case today in the newly found Soviet coal-fields. A factory can be built near an outcrop and get its coal in carts. So production started very quickly. Of course, there were other advantages, notably the absence of landlords to bargain with. But socialism would not have been enough without science.

The two thousand Soviet field geologists played a big part in building up the resistance of the Red Army, and thus helped to save the world.

Colder than the Pole

In England and other capitalist countries some scientists work in universities, generally on "pure" science, that is to say on research which is not undertaken to solve a particular practical problem. Others work for the Government or for different firms. Their results are sometimes embodied in patents, which may be used for production. But very often their only "use" is to prevent competitors from using a process.

In the Soviet Union there is no sharp line between pure and applied science. However, if a scientist wants to research on a problem which interests him but has no immediate value, he or she is often asked also to undertake some research on a current industrial problem, just as British university professors have to do teaching as well as research. And of course all discoveries, even if they are kept secret from foreigners, are put at the disposal of an entire industry.

As compared with Britain or Germany, the Soviet Union is poor in coal resources on a basis of tons per unit area, though richer in oil. And many industrial centres are a very long way from large coalfields. Even if there were much more coal, the Soviets would not squander it as British capitalists have done. So they have paid particular attention to using coal-gas as fully as possible, and also natural gas which occurs in the oilfields.

Now if several minerals occur together they can be separated in various ways. For example, tin dioxide and gold are heavier than the quartz where they are found, and can be separated by crushing it, and washing the lighter quartz away, by making the gold combine with cyanide, and so on. But it is very hard to separate gases so long as they remain gases, though fairly easy once they are liquefied by cold. Thus neon for neon lamps boils at a lower temperature than air, and is made by liquefying air, collecting the fraction which boils last, liquefying again, collecting the last fraction again, and so on. This is quite similar to the method used for separating alcohol from water in making whisky.

Of course the cold is very intense. Solid carbon dioxide, or "dry ice," is fairly familiar. It is so cold that it injures your skin if you hold it in your hand. But it is so hot relative to liquid air that if you throw dry ice into liquid air it makes it boil. And even liquid air is hot compared with liquid hydrogen.

The methods of liquefying gases were worked out largely by Dewar in London and other British, French, Dutch, German, and American workers. But these methods were suited for a laboratory rather than a factory. They are now adapted to factory conditions to some extent. But not on the vast scale needed in the Soviet Union, where even ten years ago one plant was liquefying a million cubic feet an hour of coke oven gas.

Twelve years ago the Soviet State Planning Commission decided to start liquefying gases in a really big way. A laboratory covering thirty acres was built at Kharkov, and a smaller one at Moscow. One of the first problems dealt with was how to obtain from coke oven gas a mixture containing exactly three volumes of hydrogen to one of nitrogen, so that it could be made into ammonia by passing an electric discharge through it. For ammonia was needed for fertilizers in peace and explosives in war.

Another aim is to prepare fairly pure hydrogen, and no doubt most of the hydrogen used for the barrage balloons round Moscow and Leningrad is made in this way. But this has only been made possible by very careful research; and if you want to know, for example, the boiling points of mixtures of oxygen, nitrogen, and argon, you must read Torochesnikov and Erzova's work on the subject, published in 1940.

The process of liquefying gases has been considerably cheapened by the work of Kapitza. In the ordinary process air is highly compressed and cools itself by suddenly expanding. Kapitza saw that it would lose much more heat if it was made to do work while expanding, and designed a turbine to operate at liquid air temperatures to turn as much heat as possible into work.

Today the Soviet Union leads the world both in the theory and practice of gas liquefaction. The natural gas which comes from the ground in the oilfields is liquefied; and methane, one of its constituents, is produced in such quantities that it is used for driving motor buses. In Britain we call methane "fire damp." It is the main inflammable gas formed in coal mines, and we do not use it, but blow it out into the air for fear of explosions.

Near Moscow the most remarkable of all the developments of gas liquefaction is in progress. Some of the coal seams are near ground level and only a foot or two thick. The Soviet people do not want miners to work in such narrow seams. But they do not want to waste the coal. So they put down boreholes into the coal seams. Down one borehole is forced a mixture of steam and oxygen separated from air by liquefaction. The seam is set alight, and the gas which comes up the other borehole is collected and cooled down. Not only are tar, benzene, and so on, removed, as from our lighting gas, but carbon dioxide is condensed to "dry ice" for refrigerators. The remaining gas is used for light, heat, and power, and it is hoped in future to meet the entire needs of Moscow in this way.

It is wonderful to think what this will mean from the human angle. Instead of coalmines working under dangerous and dirty conditions, and gas-works which belch out smoke and can be smelt a mile away, there will be spotlessly clean factories with machinery controlled by a few skilled men and women. Under capitalism the worker is subordinated to the machine. Under socialism the machine becomes the servant of man, as it should be.

In the present war, whenever the Soviet troops are forced back, they have to destroy such enterprises as these, which are not merely sources of wealth to the Soviet peoples, but models for the whole human race of what can be done when the workers take over production. The Soviet peoples want comfort, cleanliness, even luxury, but they are prepared to forego all these for freedom. They realize that if they were conquered, their workers and engineers would have to toil, not to raise the standard of human life, but to forge new weapons for the Nazi war machine. Rather than do this they will face cold, poverty, and death.

And those British workers who say that the Red Army is fighting our battles today, and think that this excuses them from working all out, should remember that every Nazi advance means not only a prolongation of the war and a slaughter of thousands of innocent people, but the destruction of factories and laboratories whose achievements would have brought wealth and health not only to the workers of the Soviet Union, but of Britain.

Vavilov

Among the places which the Red Army has recaptured in its offensive from Leningrad is the little town of Pushkin, formerly called Dyetzkoye Selo, or Children's village, and before the Revolution, Tsarkoye Selo, or Emperor's village. It was one of the many estates with country houses belonging to members of the imperial family.

In 1928 I spent some time in one of these houses, which had, I was told, been given by Queen Victoria to a Grand Duke. There were no grand dukes left there, but it was full of botanists doing research among rather incongruous parquet floors, marble mantel-pieces, and gilt mouldings. They would have preferred proper sinks and laboratory benches. And if, as I suppose, the Germans have destroyed this building, some of them will probably be brought back to build one which is designed as a laboratory rather than a palace. The bigger buildings at Dyetzkoye Selo were being used for country holiday homes and sanatoria for the Leningrad children, but much of the land and some of the buildings were used for a great plant breeding institute which had its headquarters in Leningrad.

As early as 1928 it had by far the largest collection of food plants in the world. In particular, wheats and other cereals from all over the world were being grown. The more important varieties were sown every year, while others were only sown often enough to keep them alive. The collection included about 100,000 varieties of cereals, as compared with 3,000 in Professor Percival's collection of wheats at Reading, which was probably the largest outside the Soviet Union. This collection served a number of purposes. Some of the varieties were good, and samples were sent to collective farms, and in the early days to individual peasants, to try whether they suited their particular soils and climate. In this way many farms were able to find a better variety than they had used before.

Other varieties gave a poor yield at best, but they had some valuable quality, such as resistance to frost, to drought, or to some particular variety of the mould called rust. It was sometimes possible, by crossing and selective breeding, to combine these with other desirable characters. Others were merely of interest in showing curious characters, such as various types of beard (which are found in some wheats as well as in barley) or purple grains. By comparison of the variation in related species, such as wheat and barley, peas and lentils, Vavilov, who directed the plant breeding institute for some years, formulated the law of homologous variation.

Thus from a study of wheat varieties, he was able to predict the possibility of types of barley, oats, and rye, some of which were later found. Similarly one can predict that some day long-haired varieties of mice and rats will turn up, and that this character will be inherited in the same way as long hair in rabbits and guineapigs.

But from the point of view of scientific theory, at least, Vavilov's most important work was on the geographical distribution of varieties of crop plants. Wheat is much more variable in some countries than others. Let us see what this means. Most people know that maize originated in Central or North America, and was grown as a crop plant by the Mayas, Aztecs, and many "Indian" tribes. Later maize was brought to Europe, Africa, and Asia. But only a few of the many varieties were brought. There are far more varieties of maize in America than in the rest of the world, and more in Mexico than in the U.S.A., apart from those recently produced experimentally. Similarly the potato came from the Andes, and more varieties are found in Peru than anywhere else.

So a botanist, who knew no history but had collected maize and potatoes in many lands, could infer that maize started in or near Mexico, and potatoes in or near Peru. Vavilov applied this principle to the most important crop plants. He found that there were more varieties of bread wheat in Persia than the whole of Europe, and more in Afghanistan than in Persia. So he deduced that bread wheats had originated in the highlands of Afghanistan or Persia, and that some varieties had been taken down to the river valleys of India, Irak, and Uzbekistan, where they were cultivated on a great scale, and spread over much of the world.

On the other hand, macaroni wheats, and other related types which do not cross readily with bread wheats, originated in Turkey or Armenia, some barleys in Abyssinia, and so on. Agriculture did not start, as had been previously believed, in the great river valleys with complicated class societies, but in small communities living in mountain valleys and very probably classless.

The work of this institute was cut down to some extent in the years before the war, largely because the best varieties had been selected, and partly because Lysenko's invention of vernalization rendered many of them less valuable than they were before. Vavilov was shot about once a year in the American press, though he continued to communicate papers to the Academy at least up to 1942.

After the war such scientific institutes will have to be restarted; and we can hope that they will be on an even larger scale than before. Vavilov's successors should be able not only to improve the crop plants of their country, and ultimately of the whole world, but to increase our knowledge of variation, and to throw new light on the origin of agriculture, and therefore of civilization.

Soviet Scientists and Blood Transfusion

Thousands of British men and women are blood donors. Very few of them know that the methods of blood storage were largely developed in the Soviet Union. Safe blood transfusion depends on a knowledge of blood groups, which were discovered by Landsteiner (now a refugee in New York) in Vienna, and Jannsky in Prague. In the 1914–18 war a lot of blood was transfused, but always directly or after keeping for a few hours at longest. Even in 1922 Keynes claimed to have established a world's record by transfusing after twenty-seven hours' storage.

In 1927 Professor Shamov of Kharkov took blood from recently killed dogs and injected it into live dogs without harming them. He approached Professor Yudin of Moscow, who in March 1930 for the first time transferred blood from a dead to a living man. The dead man had broken his skull, the living had attempted suicide and lost much blood. He was regarded as a suitable subject for this experiment, which actually saved his life. Soon after this Yudin used blood from a corpse which had been stored for three days. By 1939 this method had been fully worked out.

All cases of sudden death in the streets or public buildings of Moscow were taken to the Sklifassovski Institute, and blood was drawn from suitable corpses. It was stored for any time up to ten days, and used, not only in Moscow but in country districts, where it was sent by aeroplane. Dr. Skundina found that the blood from people who have died a violent death does not clot, except transitorily, and therefore nothing need be added to prevent it clotting during storage. This is considered an advantage, and Yudin claims that there are fewer cases of illness in the recipients than when it is taken from living donors.

This work was clearly influenced by dialectical materialism. Many years ago T. H. Huxley and others had stressed the fact that death is not an instantaneous process, as it would be if the soul left the body at a particular moment, and that the tissues might remain alive after the body as a whole had died. But this theory never became this-sided outside the Soviet Union. In the same way Filatov of Odessa was the first to introduce grafts from the cornea, the transparent window in front of the eye, from corpses into living people, as a cure for one type of blindness. Outside the Soviet Union surgeons had waited till an eye had to be removed from another patient.

Meanwhile Bagdassarov, Briukhonenko, Balakhovski, and others worked on the storage of blood from living patients. Bagdassarov finds no changes in the blood during the first four days, and has carried out several thousand transfusions with blood stored for periods up to a fortnight, though some English doctors consider a week to be quite long enough. However, with special precautions blood has been used even after a month in the U.S.S.R. In the Soviet Union blood storage has been developed on a huge scale. In Leningrad there were 1,625 donors of blood for storage as early as 1936. One thousand and ninety-four of them were women.

The Soviet methods were used on a very large scale in the Spanish Republic during its heroic fight against fascism, and were adapted to war conditions by Dr. Jorda of Barcelona. Dr. Jorda escaped to London, and has since co-operated with British doctors. The methods now in use in Britain owe much to his experience.

Meanwhile there can be no doubt that the Soviet blood storage service has been greatly expanded to meet the needs of the war. We may hope, but cannot trust, that British doctors are following the improvements which must certainly have been made in it as the result of war experience.

Marxism and Prehistory

Opponents of socialism object to such phrases as "socialist science" or "Marxist anthropology." They say there is only one science, which is based on the study of nature, and on reverence for facts, not theories; and that if anyone's politics or philosophy make any difference to his science, he is unworthy of the name of scientist. They always seem to think that they themselves start with no prejudices.

Now it is obvious that the results of scientific investigation depend on the questions which one sets out to investigate. For example, Charles Darwin devoted some years to the study of barnacles, while his eldest son George studied the sun, moon, and stars. Naturally they got different results. But, no doubt under his father's influence, the son applied the idea of evolution to astronomy, and showed how tidal friction lengthened the day and made the moon slowly move farther away from the earth.

In the same way Marxists are particularly interested in the ordinary man and woman, and they are interested in change. If you look at a list of Soviet works on mineralogy, you will notice how many deal with the transformations which minerals undergo. The most internationally famous Soviet chemist is probably Semenov, who has worked on changes in gaseous systems, for example, in explosion motors.

The field in which Soviet workers disagree most with those of other countries is that of prehistory. In a sense all students of prehistory must be Marxists to some extent, because, apart from their bones, and bones and shells which are relics of their food, almost all our knowledge of the early human peoples comes from a study of their tools. Their art is striking, but there is not much of it. A single volume will contain reproductions of all known Western European stone age art. The peoples are classified, not by their colour, language, or religion, of which we know nothing, but by their implements. Almost everywhere we find that an age of unpolished stone tools, or paleolithic age, was succeeded by a neolithic age where stone tools were polished, and these by an age of copper and bronze tools, followed by an iron age. The few exceptions prove the rule. There was never a bronze age in New Zealand, because the Maoris used stone till the nineteenth century, when the English invaders brought iron with them.

Every serious student of the old stone age must himself learn the art of stone chipping. Only when archaeologists had done this did they find out that in palaeolithic times there were two distinct methods of chipping flints; one centred in Africa where the core of a flint was shaped into a hand axe or axe head, and one centred in Asia where the tools were flakes struck off the central core. These two cultures met in Europe.

The Marxist prehistorians of the Soviet Union have added

greatly to our knowledge of undoubted facts. One single site in the Soviet Union has yielded more scuplture of the old stone age than all Western Europe. And they have also proved that in the old stone age men lived in houses as well as caves. The most remarkable of these houses, or rather its remains, was excavated at Kostienki on the Don. It was partly underground, with stout wooden walls, and probably roofed with skins or turfs. Its occupants hunted mammoths at a time when the climate was a good deal colder than now. This particular house was 113 feet long by 18 feet wide, and had a row of nine fireplaces down its centre. This is taken to mean that nine families lived in it together. We may compare this with a typical village of the new stone age, such as that of Skara Brae in Orkney.

Here there were a number of single-roomed houses, ranging from about twenty feet to fourteen feet square. Clearly the families lived separately. The primitive unity of the tribe had been broken up. This break-up coincided with the origin of agriculture and stock-breeding, and therefore of private property other than tools and weapons. If Engels was right, the private property caused the break-up.

The neolithic peoples began to make pottery. This was a slow job before the invention of the wheel. Each pot had to be built up by hand, dried, and then baked. Soviet workers have examined the finger-prints on neolithic pottery, and say that they were made by the fingers of women. If they are right, this is the first certain evidence for division of labour. Many neolithic people also spun and wove textiles, but no one yet knows whether this was done by women only. Nor do we know whether, as in some primitive societies of today, women specialized in gardening and men in herding.

Soviet archaeologists also claim to trace the growth of social classes from funeral customs. Many palaeolithic peoples buried the dead under the floor of the house or cave, so that in a sense the family was not broken up even by death. Early neolithic people went in for great communal tombs, especially round the Mediterranean. These were sometimes caves, but sometimes underground houses with imitation door-posts, and so on, containing hundreds of skeletons. The earliest neolithic tombs in Britain were long barrows, which served as tombs for numbers of people. During the neolithic age communal burial went out of fashion. In England the long barrows were succeeded by smaller round barrows containing one or at most a few skeletons, often of warriors with bronze weapons. As bronze became commoner these were replaced by cemeteries containing numerous urns. These changes have been put down to conquests, first by bands whose chiefs were powerful, later by societies where the warriors were more or less equal.

Marxist archaeologists say that they reflect changes in productive relations. Communal burial went out with communal ownership. Individual burial in round barrows showed the beginning of societies where the wealth took the form of cattle, and the chief could afford bronze weapons. Such societies are described in the book of Genesis. As bronze weapons got cheaper, most men could afford them; and societies became more equalitarian, though not communistic. Very likely the truth lies somewhere between these views. For it is probable that the origin of class society led to wars between tribes as well as oppression within a tribe.

After the war I hope that a full account may be published in English both of the facts discovered by Soviet archaeologists and of the theories which they have based on them. But it is already clear that Marxism has had a great and fruitful influence on the study of primitive human societies.

A Banned Film

I have just watched the best scientific film that I have ever seen. Unfortunately you are very unlikely to see it. It is a film of wild life in the Kara-Koom desert, near the Aral Sea in Kazakhstan. It was shown at a recent meeting of the Zoological Society. More accurately, only two-thirds of it were shown, as the show started nearly an hour late. This hour had been mainly occupied by a very acrimonious private business meeting of the society, at which various fellows attacked the society's present administration, and accused the council of using emergency legislation to stifle criticism of itself.

The Kara-Koom is a sandy desert with shifting dunes, like some parts of the Libyan desert. It is worth noting that most of the world's desert area is not particularly sandy, but consists of rocks or clay. However, sandy deserts are more picturesque, and give photographers a better chance. A certain number of willows and other shrubs grow in the Kara-Koom sandhills, and a few animals live on them. But none of the vegetarian animals are much larger than a rat, and the larger ones are carnivores. As some of these were shown catching their prey, bodies such as the L.C.C. will not license the film for public exhibition.

The first character to appear was a little burrowing rodent something like a squirrel, which came out of its hole to nibble the plants. The technical level of the film was amazing. Presumably it was taken with a telephoto lens with a camouflaged camera, but the detail was given with a softness which would have done full justice to any actress' complexion.

An owl was seen on the look-out for its prey, but it was not till a good deal later in the film that this ground squirrel, or another similar one, met its death. It was killed by a snake related to the boa, which burrowed just below the surface of the sand, so that all that could be seen of it was a ridge of sand growing at the front. This boa rolled itself round its prey, and squashed it to death very quickly. It then licked it, and expanded its jaws as only a snake can, to swallow it. By comparison with the snake's head this was about as serious a task as for a man to swallow a football. But an animal which only gets a meal a month cannot aspire to elegant table manners. In the next shot, the snake, or another of the same species, was attacked by a large lizard, and defended itself by wrapping itself round its head and neck. After several rounds both reptiles had had enough, and separated, though I think the snake, which had several bites in its neck, had had the worst of it.

We then went on to a scene of primary accumulation by force which would have delighted Herr Dühring. An industrious dung beetle had formed a ball of dry dung which she was rolling along to a suitable place for burial. She was attacked by three other members of her species, and there was a fine confused fight. As beetles are heavily armoured none of them was hurt, and finally one of them rushed off with the ball, dug a hole for it, laid an egg in it, and covered it with sand.

The next character was an animal of a kind somewhere between a spider and a scorpion, reputed to be poisonous to man, and not found in England. This was seen killing a scorpion and a lizard, and fighting with its own kind.

Finally we watched a poisonous snake. Instead of moving head first like every other vertebrate of which I can think, it moved sideways like a crab, with a loop of its body in front. Thus when the loop had reached any point, it could instantly fling its head several feet in front of it. In this way it killed a small jerboa, or jumping rat, which died in convulsions in about half a minute, and was eaten. As I have had several convulsions myself as the result of poisoning, though not with snake venom, I have no reason to think the jerboa suffered any pain.

Unfortunately, owing to the internal conflicts of the Zoological Society, we were unable to see the third reel of this film, which I am told showed the irrigation of this desert, and ended on a peaceful note.

Ought such a film to be shown, or is it the modern equivalent of the fights between animals which the ancient Romans watched in the arena? Certainly it included nothing as horrible as that very familiar sight, a cat playing with a mangled and dying mouse. These wild animals wasted no time in killing and eating their prey. We are much too apt to sentimentalize about animals, and forget that beasts are apt to be beastly. Many wild animals live by killing others, and any account of their life which glosses over this fact is simply false.

I believe that this film, quite apart from its great artistic merit, is of quite sufficient value to warrant showing it to students of zoology both at universities and schools. And I find it is a little hard to believe that an audience, which had walked to a cinema through bombed streets, would faint at a record of the fact that animals, as well as men, kill one another.

Genetics in the Soviet Union

It is somewhat difficult to get an objective view of the state of science in the Soviet Union. On the one hand, along with genuine records of fine achievements, exaggerated stories of Soviet discovery and invention are put about. Typical communist success stories, did you say? We Europeans are often amused to read newspaper cables from the U.S.A. claiming credit for American discoveries which had actually been made in Europe some years earlier. And no doubt Americans have their laughs at similar stories from capitalist Europe.

As against this, we are told that science is at a very low ebb in the Soviet Union. No research is encouraged except what is thought to be of immediate value to industry, agriculture, or war. No theory may be published which does not conform with the canons of dialectical materialism. The intellectual liberty which is an essential condition of scientific progress is completely absent. And so on. One of the most important and successful lines of German propaganda in preparation for the present war was the spreading of such views as the above, with the object of preventing any co-operation of the British and French ruling classes and the Soviet Union which could have prevented the outbreak of the war.

As a matter of fact some branches of science are highly developed in the U.S.S.R., and others rather poorly. Thus physical chemistry is making great strides. Semenov's work on gas reactions is of the first importance. On the other hand, research along the lines of classical organic chemistry is less important,[1] in spite of the good work of pre-revolutionary Russian chemists such as Reformatsky. In mathematics very little is being done on such favourite American topics as finite group theory, but in the study of probability the Soviet Union seems to be ahead of America. It is easy, for propaganda purposes on either side, to pick on the bright or dark patches. In a general way Russian science re-

---

[1] About 10 per cent of the papers on organic chemistry cited in *British Chemical Abstracts* in 1939 were from Soviet laboratories, as compared with over 20 per cent of those on mineral and soil chemistry.

sembles American science forty years ago. Many of the leaders are training students in a number of different subjects rather than concentrating on one line of research. So many new institutions are being opened that a larger number of second-rate men and women are obtaining posts than in England before the war, or America today, where expansion is or was much less rapid. We may look for a gigantic flowering of Soviet science in another generation, corresponding to that of America in the last fifteen years, but on a considerably larger scale, since the opportunities for education are more widespread.

Nevertheless, even today the Soviet Union is leading the world in certain branches of science. In geography the Soviet arctic explorers have taken the lead which was held by such men as Peary and Amundsen. In cryology (the study of cold) Soviet scientists are ahead of the rest of the world in methods of separating gases from mixtures by liquefaction and fractional evaporation. Their work on soils and their transformation is superior to that of other countries, though here it must be admitted that Glinka laid the foundations before the revolution; and so in many other branches. In the rest of this article I shall deal with Soviet genetics, my own branch of science, of which I naturally know most.

Let us begin with the criticisms which have been made. Two first-rate Russian geneticists have refused to return to their country and are occupying positions elsewhere, Dobzhansky in Pasadena and Timofeeff-Ressovsky in Berlin. In the Soviet Union Tsetverikov, Agol, and Levit have lost their posts. Agol is alleged to have been imprisoned, or even executed. And Lysenko, who is admitted to be a first-rate plant physiologist, has attacked the basic theories of genetics.

Now let us look at the credit side. Under the guidance of Vavilov an immense mass of data on the genetics of cultivated plants has been accumulated. His school has also studied the related wild plants not only in the Soviet Union, but as far away as Abyssinia and Peru. This work has led to some very important results. Vavilov was the first to formulate the law of homologous variation in related species, now confirmed and extended by Sturtevant and other workers in America. He determined the places of origin of our more important cultivated plants. This was done under the direct stimulus of Marxist theory, according to which the domestication of these plants was a far more important historical (or rather prehistorical) event than the wars and other political happenings with which written history is mainly concerned. Special attention was paid to the evolution of weeds. These may evolve into cultivated plants. Thus rye is a weed in the wheat crop in warm climates, forms a mixed crop with wheat in primitive agriculture at intermediate temperatures, and replaces wheat in the north or on mountains.

A vast amount of detailed observation of plant chromosomes was done by Levitsky, Navashin, and others. This was necessary for Vavilov's work, and has put the whole question of crop plant hybridization on a more scientific basis. A number of very remarkable hybrids, for example, between wheat and couch grass, are now being tested out.

In the field of fruit genetics we may notice Rybin's synthesis of the plum from the hybridization of the wild cherry-plum and sloe. There can be little doubt that our cultivated plums originated in this way. On the other hand, Soviet geneticists have done little or nothing on the genetics of ornamental plants such as the sweet pea, the poppy, and the various *Primulas*, which have led to important theoretical results elsewhere. They have concentrated on economically important plants, though their studies of them have been very thorough, and have included problems of no immediate economic importance.

In animal genetics Soviet workers on poultry such as Serebrovsky have covered much the same field as those of other lands; but as regards sheep, cattle, camels, and other larger animals, they are in a class by themselves. For example, Vassin is now mapping the genes on sheep chromosomes. The large scale of Soviet animal husbandry makes artificial insemination on a vast scale possible. A single ram or bull may have several thousand children available for study. A particularly interesting line is the study of the biochemical differences between and within breeds. For example, the blood chemistry of race-horses and cart-horses is compared, and also that of efficient and inefficient members of the two breeds. Nothing of this kind is being done elsewhere.

"Formal genetics," as it is called in Russia, received a great impetus from the visits of C. B. Bridges and H. J. Muller, two leading American geneticists, who introduced *Drosophila* to the Soviet Union. This little fly gets through thirty or more generations a year, and you can grow four hundred in a milk bottle, so it is uniquely suited for the study of inheritance. Russian workers took it up enthusiastically, but much of their work was inspired by Muller, and was of the same general character as similar work done in the U.S.A. However, one group took up the genetical analysis of populations, which had been started by Soviet poultry and cattle geneticists, and applied it to *Drosophila* populations. It turns out that although the flies look alike, large numbers of them carry concealed recessive genes. So when their offspring are inbred, a great variety of abnormal insects is produced. This line of work was started by Tsetverikov, but carried on on a vast scale by Dubinin and others. It has been confirmed on a smaller scale in the U.S.A. and Britain, and has led to new perspectives both of evolution and of human congenital disease.

Let us now look at the criticisms against this background of solid and often brilliant achievement. Dobzhansky and Timofeeff-Ressovsky got good jobs abroad, as dozens of British scientists have done in the last twenty years without any suggestion that British science is persecuted. Tsetverikov was a serious loss to research. The other two dismissed workers had not done work of great originality. But several good British geneticists have recently lost their posts, one for marrying a Chinese wife, another for trying to expose corruption in an institute, and a third for disproving one of his professor's pet theories. Similar events have occurred in America.

Lysenko's attack on genetics is much more interesting. The public in the Soviet Union is intensely interested in biological problems, and Lysenko's attacks were widely reported in the daily newspapers. Now such attacks are not uncommon. Professor Jeffrey of Harvard has attacked genetical science much less temperately and on much flimsier evidence than Lysenko. So has Professor MacBride in London. But such attacks are not hot news in New York or London, because the publics of those cities are much less interested in genetics than is that of Moscow. Some of

Lysenko's points are, I think, valid against genetics as often taught, rather than against the theories held by competent geneticists. He was quite right in saying that so-called pure lines of plants are generally mixtures, and that an exact three to one ratio in accordance with Mendel's law is very rarely obtained. He also stated that in tomatoes and related plants a number of characters described as hereditary can be propagated by grafting. In just the same way Little, Bittner, and other workers at Bar Harbor, Maine, found that the tendency to breast cancer in mice, formerly regarded as hereditary, was largely transmitted through the mother's milk. Lysenko further pointed out that a great deal of successful animal and plant breeding is carried out without any reference to the results of genetical research in the last forty years, and that geneticists have made exaggerated claims for the economic value of their science. In both cases he was right, though the economic value of genetics is greater than he thinks.

I am convinced that he went much too far both in his attack on the chromosome theory, and in his claims concerning the possibility of transferring characters by grafting. But what has been the result of his attacks? Vavilov was their chief target. Vavilov still directs research on a vast scale. So far from having been muzzled for his alleged anti-Darwinian views he communicated seventeen papers on genetical topics to the Moscow Academy of Sciences between January 1st and April 10th of 1940.[1] Lysenko attacked "formal genetics," that is to say genetics which is concerned with such questions as locating genes in chromosomes, rather than in finding out how they act in the development of an individual, or arise and spread during the evolution of a species. It may be that under the stimulus of so brilliant a teacher as Muller, an unduly large fraction of the younger Soviet geneticists had occupied themselves with formal genetics. However that may be, formal genetics goes on in the Soviet Union, and the output of work in this field is a good deal larger than in England, even before the war.

In the controversy between Vavilov and Lysenko, I would personally give Vavilov best on most points. Nevertheless, I

[1] Vavilov's name is now less prominent, but up till June 1941 the output of genetical work showed no sign of abatement.

welcome the controversy, and wish that similar debates elsewhere were given equal publicity. I have little doubt that when I
taught genetics (owing to the war I no longer do so) I made a
number of misleading statements. I should be a better teacher if
these were pointed out in a public debate to which I could reply.
But in England things are done differently. Five years ago there
were two professors of genetics in England. Now there is none.[1]
These chairs were not suppressed as the consequence of a public
debate, but in all probability as a result of some old gentlemen
talking the matter over privately after a good dinner. If my
science must be attacked, I prefer the democratic Soviet method.

I think the position of genetics is fairly typical of that of Soviet
science in general. Large-scale work, so far as possible, is concentrated on organisms, substances, or processes, which may be of
economic importance, but a great deal of latitude is allowed. Any
knowledge about cows, coal, gas explosions, or arctic ice, may
be of value some day. So there is no restriction on what aspect is
investigated. If basic principles can only be worked out on economically unimportant objects such as *Drosophila*, then these are
used. In all research the historical angle is stressed so far as
possible, whether it be a question of human history as in the case
of Vavilov's work on crops, or of changes in insect populations,
as in Dubinin's. This latter tendency, along with a distrust of
over-mechanical theories, is no doubt an effect of dialectical
materialism, and to my mind a desirable one.

But as dialectical materialism is a method of thought and
action, not a dogma, it is hard to see how it could influence
decisions on such controversies as this, except indeed by sug-
gesting that certain possibilities should be explored, even were
every Soviet scientist compelled to adhere to this philosophy,
which is, of course, not the case. Anyone who studies the record
of the genetical controversy recently published in *Under the
banner of Marxism*, and particularly of the interruptions, will
certainly realize that thought on scientific topics is pretty free in
Moscow.

I must confess that the genetical theory of racial inequality,
widely held not only in Germany but in the U.S.A. and Britain,

[1] One chair has since been revived.

which has played its part in bringing about such events, seems to me considerably more important than those which are now being disputed in the U.S.S.R. And I could wish that those of my European and American colleagues who have taken up the cudgels on behalf of Vavilov, who is not incapable of self-defence, would transfer some of their energies to an attack on this doctrine.

Soviet Children as Scientists

The children of the Soviet Union are fighting heroically in the occupied regions, and elsewhere they are taking their full share in production. I am going to write about some of the things they did in peace time, and will do so again when the war is over.

In Britain many school children learn science, but they have little chance of making any discoveries for themselves. In the Soviet Union some children make discoveries. Here are two examples of how country boys and girls worked on birds.

As in our own country, many birds fly south in autumn to warmer countries, and north in spring. British birds have to cross the sea, but in the Soviet Union they fly thousands of miles overland. In order to discover what routes they took the children in hundreds of country schools started trapping birds. The traps were carefully designed so as not to hurt them.

Each bird had a numbered ring put on its leg and was then let loose. Results began to come in when the same bird was caught two or three times in different places. The dates and places at which birds were caught were noted down, and the results worked out by other school children in the big towns.

It was rather like the way in which Fighter Command Headquarters follows the movement of German bombers from reports of the Observer Corps, except that of course there was no hurry about it. The work was not finished when war broke out. But when it is complete, a good deal more will be known about the migration of Soviet birds than about British.

Other children, or perhaps the same ones, studied the feeding habits of starlings and other birds which can be induced to nest in a box. When the nestlings were fairly well grown, they were taken out of the nest, and replaced by a wooden figure of a nestling which opened its beak when the mother or father alighted on a perch to feed it. The children made the dummy nestlings themselves, though I don't suppose they designed them.

The birds, or at least some of them, were completely taken in, and kept on feeding the dummies. The food collected in a tin below. The next job was to take it out, and count the numbers of different kinds of insects, worms, and so on, collected each day.

In this way the children could learn for themselves which birds were most useful to the farmers and gardeners in keeping down pests. The same kind of work has been done in other countries, but it generally involved killing a bird to see what food it had in its crop. And if research of this kind is done in hundreds of schools, it must give a mass of information which will not only help agriculture, but tell us a lot about how birds adapt themselves to different diets in different surroundings.

These are the children whom the Nazis regard as uncivilized members of an inferior race, who must be massacred to protect Europe from the menace of Bolshevism.

Blubo

If I can follow the Nazi philosophy (and I may well have failed in this very difficult task) the characteristics of a race depend on its blood and soil. This doctrine of "Blut und Boden," or "Blubo," as it is called for short by the children who have to attend lectures on it, has a primary appeal to the emotions. Nevertheless, it claims some sort of scientific basis, just as numerous theologies, which represent illegitimate intellectualizations of the quite genuine religious emotion, have claimed to be queens of the sciences in the past.

I am no pedologist, as those who study soil call themselves. But my meagre acquaintance with that science leads me to believe that while the soils of Germany are very diverse, none of them is peculiar to Germany. Friesland is not unlike northern Holland, Brandenburg is like Western Poland, and so on.

With haematology, on the other hand, I have more than a bowing acquaintance. I spent three months in learning to measure amounts of oxygen and carbon dioxide in a cubic centimetre of blood. As a hospital biochemist I have performed estimations of blood urea to decide whether or not it was safe to operate in cases of prostate disease, and many other routine analyses. I have certainly produced and measured greater changes in my own blood than anyone else has ever done. And it is about eighteen years since I first learned how to determine the group to which the blood of an individual belongs. So perhaps my knowledge of human blood from a scientific point of view is equal to that of Herr Hitler from the emotional angle.

And the scientific study of the connexion between blood and race has led to very definite results. The pioneers were two Polish doctors, Ludwik Hirzfeld and his wife Anna. They worked at the State Institute of Hygiene in Warsaw, so they may be dead today. But their work is not.

In 1900 Landsteiner in Vienna discovered that transfusion of blood from one man or woman to another caused illness or death in certain cases, and that the bloods which he investigated fell into three groups. He did some preliminary work on their inheritance. In 1907 Jannsky in Bohemia discovered a fourth group, and Moss in New York confirmed his discovery independently in the next year. During the war of 1914–18 blood transfusion became important, and it was in Salonika, where the Hirzfelds were working with the allied armies, that they made the striking discovery that the frequencies of the groups were very different in different peoples. They sent a paper on their discovery to the *British Medical Journal*—which refused to print it. Finally it was published in the *Lancet*.

In 1910 von Dungern and Hirzfeld first systematically investigated the inheritance of blood group membership, but it was not till 1924 that Bernstein of Göttingen, now, like Landsteiner, a refugee in New York, stated the laws of this inheritance in the form which is now almost universally accepted as correct. These laws are used, under the Bastardy Act in Britain, and similar laws in other countries, as tests of paternity in disputed cases.

The full significance of the Hirzfeld's discovery now began to appear. Physical anthropologists had long been measuring physical characters, such as skin colour and skull shape in different human groups. They were known to be partly determined by heredity, but environment certainly influences them to some extent. Our skins become browner in summer. Certain primitive folk distort the heads of their children as civilized nations distort their feet. On the other hand blood group membership is absolutely fixed at birth (and indeed much earlier) and the laws of its inheritance are extremely simple. For the first time anthropologists could study a character in whose genesis environment played no part. If similar psychological characteristics are ever found, the problem of the relation between race and culture will be soluble, at least in principle.

It would have satisfied those who stress the difference between human races if the distinction between races had been sharp. If you take a lock of hair from a European and a negro you can assign it to the right race with almost complete certainty. But if you take a drop of blood you cannot. Actually if the bloods are taken from an Englishman and a Senegalese respectively, you will get the right answer in 62 per cent of cases, instead of 50 per cent by mere guessing. In the most favourable case, the distinction between an Eskimo and a Blackfoot Indian from the Rocky Mountains, you would only be right in 82 per cent of cases. If you used other more delicate techniques you would increase the certainty of your diagnosis, but you would never be right in 100 per cent of cases. In fact, if you want to be accurate it is better to say that you are of pure European hair than of pure European blood!

The reason for this is that every race so far studied includes members of more than one group, and most of them include members of all four. So the characteristic of a race is not membership of a particular blood group, but the proportions in which the various groups are found. The four groups are sometimes numbered, but are more usually designated as O, A, B, and AB, according as the blood corpuscles carry neither, one, or both, of two substances A and B. No one has been able to demonstrate any increase or decrease of fitness due to either of these substances. Hence the frequencies remain almost constant from one generation to another. Nor do they change, at least within a few centuries, when a people migrates to a different soil and climate. On the other hand, when two races intermarry the proportions in the mixed race are intermediate.

There is, of course, nothing in all this to surprise physical anthropologists. When we say that the Swedes are long-headed and the Swiss round-headed, we do not mean that every Swede has a long head and every Swiss a round head. We mean that long heads are commoner in Sweden than Switzerland. Blood groups are a better anthropological character than head shapes in so far as they are clear cut and their heredity is understood, but a worse one in so far as they cannot be determined on skeletons, though this is occasionally possible with mummies.

Germany and Japan have led the world in the determination of blood group membership on large numbers, but good data exist for Finland, Poland, and some parts of Italy, whilst the pre-Columbian peoples of America have been extensively studied. The Soviet data are rather scrappy considering the vast material available, but are better than those for Britain, France, Spain, Sweden, and Norway. Over a quarter of a million determinations have been made in Japan, as compared with less than a thousand in Scotland, Ireland, or Wales. When, however, the data on those who have volunteered as blood donors during the present war are collected, the British figures should be fairly satisfactory,[1] though still not so good as the German or Japanese. Steffan and Wellisch in Germany and Boyd in Boston have collected most of the data for the whole world. The differences are obvious when samples are large. Thus among about 5,000 Londoners, 45 per cent belong to group O, 44 per cent to group A, 8 per cent to group B, and 3 per cent to group AB; among 30,000 Berliners the figures are 37, 41, 15, and 7 per cent, whilst 5,000 inhabitants of Leningrad have 32, 37, 23, and 8 per cent. The most striking change as we go eastwards in Europe is the increase in the number of people who have the B substance, from about 10 per cent in Britain, Spain, and Belgium, to 37 per cent in the Perm district of the Urals.

At the International Genetical Congress held in Edinburgh in

[1] They already are so in 1944.

Europe I showed maps based on counts of seventy-five European populations. We can draw contour lines across Europe so that to the east of a given line, all large samples have more than, say, 20 per cent of members carrying the B substance, while to the west of it less than 20 per cent possess it. These contour lines run roughly north and south, though of course they are not quite straight. For example, the Czechs have more of the B substance than the German-speaking peoples to their north and south, and the Greeks and Turks have less of it than the peoples of the Balkans to the north.

If we made a relief map on the basis of these contours, we should find that the Soviet Union was represented by a large plateau with a gentle slope, broken by a few peaks caused by primitive folks such as the Votyaks, and depressions represented by Jews. Curiously enough, the blood group frequencies among the Jews of Odessa are very close to those of the Gentiles of Königsberg. Esthonia, Latvia, and Poland form part of the Russian plateau, but Finland and Lithuania are well below it, though above Germany.

The contour lines are crowded together in the Baltic, since Scandinavia has far less of the B substance than the countries east of that sea. There is also a sharp slope from east to west in Germany. East Prussia is on the level of Bulgaria and the Swedish-speaking Finns, whilst Munich and the Rhineland are comparable with Scandinavia, Scotland, Paris, and Lombardy. In western Europe the level of B is low, but less regular than that of the Russian plain. The lowest figures for B are a rather doubtful one for Madrid and a still more doubtful one based on a small number of Basques.

In Eurasia as a whole we find the highest proportion of B among the "depressed classes" in India, though even among Brahmins it is well above the European level. It is also high among the Buriats in southern Siberia. The frequency falls off as we go either eastwards or westwards from central Asia into populations where it was originally rare or absent.

Among the Atlantic peoples who have little B, there are great differences in the amount of A. The people of Ireland (both Dublin and Belfast) have far less of it than those of London and

East Anglia; and the scanty data for Scotland, Wales, and western England, seem to put them in an intermediate position.[1] Germany, because its forests and mountains acted as barriers against infiltration from the east, shows such heterogeneity that one cannot speak of a German race as one can of a Japanese or even a Russian race. But the British Isles, which seem to include samples of the neolithic peoples of western Europe little affected by later movements, are even more heterogeneous than Germany.

To take a single example of the application of "Blubo," the bloods of the peoples of Danzig are intermediate between those of eastern Germany and Poland. So on this criterion Danzig should remain a "free city." To my mind such a conclusion deserves to be ranked with the mitigation of the Sunchild's sentence in Erewhon on account of the meritorious colour of His hair. It seems fairly obvious that the primary criterion of the legal nationality of any people should be their own wishes, while the economic and strategical interests of neighbours should not always be neglected.

The facts about human blood form a part of the science of physical anthropology, as do those concerning skin pigmentation. If the pigments in the skin of our Indian fellow-subjects absorbed in the ultra-violet region only, instead of the visible region, as they actually do, we should be unable to distinguish them except by careful tests, and should be much more likely to treat them as equals. We may hope that, by the time the facts concerning blood group membership are widely known, it will also be realized that they serve to unite the human peoples rather than to divide them. If I want a blood transfusion, and no tests are available for the donors, I shall do best, as a member of group O, to pick a Maya or a Pueblo Indian. If you belong to group B you had better choose an "untouchable" from India. Perhaps it might have been better for the Nazi philosophy if Hitler had confined his attention to hair and noses, and left the study of blood to those who are more interested in saving lives than in taking them.

[1] This is confirmed by figures published during the war.

Nazi Lessons in British Schools

I was recently asked to address a number of London school teachers on the teaching of biology. As a preliminary I read a number of the publications of the School Nature Study Union. Many of them were excellent.

The suggested lessons were often very well designed to interest children in the animals and plants around them. But the attempts to link this interest up with everyday life were not so satisfactory.

The worst example occurs in a Scheme for the teaching of biology drawn up by a small group (I am glad it is small) of H.M. Inspectors of Schools, and published in the School Nature Study Union's publication Sixty-seven. Lest I should seem unfair I quote several sentences.

"If the children could really become interested in the various ways in which living creatures show intelligence, and in the particular habits and modes of life which are associated with race in general, they might be in a fit state of mind to appreciate the truth that each of the various races of man has its own mode of life and thought."

The writers go on to ask "if knowledge of this kind is not indispensable to the race that rules an Empire."

Hitler thought that beliefs of this kind were indispensable to the German "race," which he believed was destined to rule an Empire including our own country. They are meat and drink to anti-semites. They may be useful to British imperialists, too. But why call them "truth"? Certainly it is a good thing for children to realize that other peoples have very different customs from our own, and that men and women can lead good and useful lives even though they wear no clothes and eat raw meat.

It is true that racial differences are often associated with different modes of life and thought. The Eskimos are racially different from the people of Jamaica, and it would be a miracle if a people spearing seals among the ice did not live and think differently from one growing bananas in a hot climate. But do these differences of life depend on differences of race? If so they ought to remain constant unless the racial composition is altered. And according to the Nazis they do.

Let us look at a few examples. Eleven hundred years ago the Danes were the most warlike and bloodthirsty people in Europe, perhaps in the world. They had conquered half England, and part of France. They raided as far as the Mediterranean. It was a disgrace for a male Dane to die anyhow except in battle. Since the time of the Vikings there has been much less immigration into Denmark than into Britain, France, or Germany. The race is the same. But the modern Danes have offered less resistance to the Nazis than any of the other conquered peoples. To my mind they are as much too gentle as their ancestors were too fierce. Which is the Danes' "own mode of life and thought," scouring the seas in search of plunder and a glorious death, or growing food and making munitions for their conquerors?

Again, the "Red Indians" of North America were much alike in their physical characters, and on any possible classification formed one race. Most of them were hunters, and fought with great courage and great cruelty, collecting the scalps of their enemies and roasting prisoners alive. But in Arizona men and women of the same race were agriculturists, living a peaceful life in towns called Pueblos. The fierce hunters have been liquidated and their descendants have changed their customs, but the Pueblo "Indians" continue to live much as their ancestors did. The difference in life and thought was not due to difference of race, but of production.

Finally, think of the modern Americans. They have their own modes of life and thought, but these are much the same for Americans of English, Irish, Scandinavian, Italian, or Jewish ancestry. The coloured people, with their history of slavery and persecution, have a rather different way of life. But even this is essentially American. Americanism does not depend on race, but on nationality and all that this means. There may be an American race in the future. Today men and women of many races share a common culture. This, by the way, is vastly superior to the version of it with which we are familiar in this country, thanks to the Hays film censorship.

Missionaries have done a good deal of harm to some primitive peoples. But at least they have always believed that men and women of every race could adopt Christian "modes of life and thought." And in consequence the best of them have done much to protect "savages" against the worst features of imperialism. The racial theory of the school inspectors is as flatly opposed to Christianity as to communism.

Actually if the Inspectors knew a little more biology they would know that differences of habit are mainly associated with differences between species. The thrush and blackbird, with their different songs and habits, are different species, though closely related, and do not interbreed, as human beings of different colours do when they inhabit the same country.

The races, or breeds, of dog, certainly differ in their behaviour, and the differences are hereditary. But this is because they have been selectively bred. But we do not mate human musicians to musicians, and drown the children who are tone-deaf. So there are no human breeds with special inborn tendencies like the different dog breeds.

On the other hand, important differences in habit within a species are often quite independent of differences in colour or structure. Thus some British starlings migrate, and some stay here all the year round. The difference seems to be inherited, but both types look alike.

It is true that a group of insects which were originally classified as one species is sometimes now divided into "biological races." For example, mosquito "species" are now split up according to what animals they bite, how they lay their eggs, and so on. But as the different "races" do not give fertile hybrids, they are far more different biologically than the races of mankind, and could be described as different species.

But in spite of these facts, racial theories very like Hitler's are doled out to children, at least by teachers who want to please certain inspectors. They are most useful to imperialists. Kipling was full of them. "East is east, and west is west, and never the twain shall meet" is an excellent slogan if you don't want the Indian people to enjoy representative government.

At the present time we are fighting Hitlerism with bombs and depth charges. We should be doing so in the realm of ideas also. I do not mean that we should condemn every idea which the Nazis have ever held. No human being can be 100 per cent wrong on all questions. But there is absolutely no excuse for teaching Nazi theories to children when these have no basis in fact, and I hope that those who realize how much our future depends on what is taught to children will take up the matter, and prevent the teaching of bogus biology which can be used as the theoretical basis of British fascism.

Race Theory and Vansittartism

The Nazis teach, and many of them believe, that the Germans are a race different from all others, and possessing inborn qualities which make them fit to rule other races. This is a convenient doctrine for the German armament millionaires, the general staff, and the Nazi bosses. It was not so good for the Germans who died in the snow at Stalingrad.

It is now beginning to recoil on the Germans like their technique of bombing civilians. Many people believe that the Germans are a different race, and have a specially aggressive and brutal nature, which is inborn in them. This is in fact the theory behind Vansittartism. If it is true the only thing to do with the Germans is to massacre them or keep them in perpetual bondage.

But is it true? Anthropologists classify men by various physical characters, such as the colour of their skin and eyes, and the shape of their hair and skulls. These characters are inherited, though all except perhaps eye colour can be influenced by environment. Englishmen can get sun-tanned, and negro hair can be straightened. On the basis of such characters one can distinguish the main races of mankind with certainty. One can separate men of European, Chinese, or negro stock from one another, even if the Europeans have lived in Africa or the negroes in America for some generations.

Of course we must be very careful not to confuse race and nationality. Thus Mr. Learie Constantine, the famous negro cricketer, both of whose parents were British born, has quite as good a claim to British nationality as Mr. Churchill, whose mother was American born. But not all, though probably some, of Mr. Constantine's ancestors were racially British. Unfortunately we use the same word for race, language, nationality, and sometimes religion. This confusion can be used to support all sorts of injustice. Hitler forced German nationality on the Austrians because they spoke German. A hotel manager denied Mr. Constantine his rights as a British national because his skin is dark.

The Germans have no claim at all to be a race on the basis of inherited characters. In parts of northern Germany the commonest type is tall, fair, blue-eyed and long-headed, like the typical Swede. In East Prussia a tall, fair type with high cheekbones is found, as in Russia. In the south the Germans tend to be short, with brown hair, round heads, and a tendency to horizontal rolls of fat on the neck. The same type is common in Austria and Switzerland.

In fact if Europe were divided up on a basis of inherited physical types, parts of Germany would be united with Scandinavia, other parts with Switzerland, and so on. Nevertheless, it is true that some mental characteristics are very common in Germany, some of them being very undesirable. In behaviour the Bavarians are more like Prussians than Swiss, though physically they are more like the latter. The reason for this is not racial but historical. Feudalism was stronger in Germany than in any other part of Europe. Germany was not unified till 1871, and even then a number of subordinate kings, princes, and Grand Dukes survived until 1919. This survival of feudalism was possible because agriculture was far more primitive than in Britain, France, or Holland, and manufactures far less developed. In the nineteenth century the country was very rapidly industrialized and its mines developed. But much of the feudal structure remained.

There was never anything like the French, or even the English, revolution to destroy it. In 1919 such a revolution began, but it was put down with the aid of the British and French Governments, and we are now paying the price. A feudal aristocracy is naturally aggressive, but knights in armour could seldom go very far, as the economic structure of feudalism made standing armies impossible. Besides, they were vulnerable to bowmen. But a knight in a tank, supported by the whole structure of modern industry, is the very devil. The feudal landowners of Germany provide the bulk of the higher officers. In the last generation they have intermarried with the bourgeois capitalists, while the Nazis gave them enough mass support for the conquest of Europe.

As long as the class structure of Germany is preserved, the Germans will go on being aggressive, even if it is occupied by allied forces for a generation. Germany will only cease to be aggressive when its ruling class is wiped out. I do not mean that they should all be massacred, though I hope that those who are actually responsible for murders will be killed. I mean that they should cease to exist as a class, as they will do if they are deprived of the surplus values on which they live at the expense of other Germans. This could be accomplished in several ways, of which I should prefer to see the method of the Russian Revolution adopted. But provided the landlords go, it may not very much matter to the rest of the world whether the land is divided into small holdings, collectivized, or nationalized. Provided monopoly capitalism goes, the type of socialism adopted is of more importance for the Germans than for the rest of us.

The question which Marxists should ask followers of Vansittart is this: "Do you stand for the liquidation of the German ruling class?" If they do not, they are merely preparing the ground for further German aggression, and no amount of re-education or military occupation will prevent another war.

What to Do with German Science

The main object of the occupation of Germany will be to prevent the Germans from preparing for another war. This will involve a control of German industry so as to make the manufacture of armaments impossible. But a difficulty at once arises.

The countries which have been invaded, such as France and Poland, will need machinery to build up their own economies. For example, the Polish peasants will need tractors if they are to raise their agricultural productivity to the level of the Soviet Union.

But a factory which makes tractors can easily switch over to making tanks. No doubt this problem has already been considered. Probably a committee at Yalta was dealing with it. But the problem of German science has not been seriously discussed. A number of eminent British scientists who enjoy the full confidence of the Government have no knowledge of any definite plans on this matter.

A certain number of German scientists have been guilty of murder. They have organized the mass killing by chemical methods of Jews and other persons judged to be of inferior race, and are alleged to have experimented on them with various drugs. They must be hanged like any other murderers.

I don't object to experiments on men and women, even dangerous ones. I have done them. But I have always been one of the subjects. Indeed I have never done any possibly painful experiments on animals which I have not also done on myself, if only because one does not know as much as possible about an abnormal physiological condition till one knows what it feels like.

Most German scientists have done nothing worse than the majority of German civilians. But some of them are more dangerous. There are those in this country who would like to stop all German research work. I think they are wrong for three reasons.

In the first place a great deal of German research, even in the last twelve years, has been of benefit to the whole of humanity. For example, the first of the series of sulphanilamide derivatives, such as sulphapyridine and sulphathiazole, which have proved so valuable in many diseases, was produced by a German, Domagk.

Secondly, we must hope that the Germans will ultimately take their place among the civilized peoples. They cannot do this without intellectual culture, which includes science. To take only one example, unless they learn some real biology they will never understand the utter falsity of Hitler's racial theories.

Thirdly, it is not realized how long it takes to put a new discovery in fundamental science into practice. Suppose I discover tomorrow that by injecting suitable hormones into a sheep I can double its wool production per year, this does not mean that the wool production of England will be doubled in ten years, or would be doubled even if there were no vested interests in the way. Probably the hormones could only be got from some organs in dead animals, and the process of preparing them would be difficult even in the laboratory. To produce enough for half the sheep in England it might be necessary to find out how to make them from coal tar derivatives, which would be a problem for organic chemists.

The first radio messages were sent exactly fifty years ago. I still remember the excitement when the murderer Crippen was arrested because the captain of the ship on which he was fleeing had actually picked up a radio-telegram. That was in 1910, after fifteen years of development.

No research should be allowed in Germany without a licence, certain types of applied science being absolutely barred, and all places of research should be open to inspection without notice, as physiological laboratories are in England.

If the allied control is genuine, this should be quite sufficient to prevent research directed towards a future war. If the control is not genuine, as it was not after 1918, then more important war preparations than research will go on.

A more difficult question is the future of the German scientific industry. A factory which makes microscopes and telescopes can easily turn over to making gun-sights and bomb-aimers. But the Germans have deliberately destroyed and plundered scientific institutions in most of Europe, and our own have been unable to buy badly needed instruments since 1940.

American optical firms are busy on war work, and will be till Japan is conquered; British firms and Soviet factories are still busier. Unless such firms as Zeiss are made to start making scientific equipment once more, the rebirth of science all over the world will be held up for years, and thousands of people will die of preventible diseases.

To take one example, malaria is a serious cause of death in Yugoslavia and Greece. It is carried by mosquitoes. To identify mosquito species one needs a low-powered microscope. Thousands of microscopes will be needed to set up an adequate anti-malarial service in south-eastern Europe.

Of course competitors in Britain and the U.S.A. would be very glad if the German optical factories were liquidated. The World Trade Union Conference, I am glad to see, have come out "for the utilization within the limits imposed, of German industrial and other resources for the rehabilitation of countries which the Germans have devastated."

The key to the whole problem is the completeness of the control which is to be exercised. It would certainly be better to destroy the Zeiss optical factory than to allow it to make enough microscopes to pay its way, while building up a skeleton organization for war work. But it would be better yet to use it to make the microscopes which are needed even in Britain, but still more in the formerly occupied countries.

The world needs German science and the technical skill of German workers. They can only be directed into safe channels if the structure of German capitalism and landlordism is destroyed, and the military occupation lasts for at least twenty years. There are two alternatives to this policy. One is to relax our grip and allow a new Hitler to come into power. The other is to destroy every factory and mine in Germany, which would starve many Germans, and impoverish all Europe.

Fortunately the Soviet Union, which has suffered far more from German aggression than Britain, let alone the U.S.A., will have a large share in determining the policy adopted. And it will not choose either of the last two alternatives. Under a sane policy there will be a place for German science.

8

HUMAN LIFE AND DEATH AT HIGH PRESSURES

MEN are exposed to high pressures in a number of circumstances. They may be working in compressed air in a caisson or diving-bell, working under water in a diving dress, or attempting to escape from a sunken submarine. In the latter case it is obviously necessary that the air pressure inside a part of the ship should be equal to that of the water outside before a man emerges. This can be achieved either by flooding a small escape chamber holding only two men, or a whole compartment of the ship. Men have escaped by both these methods. They can rise through the water either holding their breath or breathing from a Davis submarine escape apparatus. The former method is not to be recommended, but it is not quite so hazardous as it sounds, for a lung-full of air at five atmospheres contains as much oxygen as a lung-full of oxygen at atmospheric pressure, and will allow a man to hold his breath for more than twice the normal time. The Davis submarine escape apparatus consists of a rubber bag and a soda-lime canister to absorb carbon dioxide. The bag is filled with oxygen from a small cylinder of the compressed gas. It has the advantage over air that it can be used almost to the last dreg. We shall come to its disadvantages later.

In June 1939, H.M. Submarine *Thetis* was sunk with civilians as well as naval officers and ratings. The Amalgamated Engineering Union and the Electrical Trades Union asked me to attend the investigation of this disaster, as some of their members had been killed. I was only able to carry out some very rough experiments during the course of this inquiry, but they made it clear that certain physiological factors concerned in escape from submarines had not been fully considered. I was therefore asked by Admiral Sir Martin Dunbar-Nasmith's physiological sub-committee on escape from submarines to undertake further research on this question, and it has very kindly permitted me to publish certain results. Messrs. Siebe Gorman and Co. put their plant and staff at my disposal. All the experiments described here were carried out in a small steel chamber at their works, which holds two, or at a pinch three, people in a sitting position. The experiments were conducted by Dr. E. M. Case and myself, on ourselves and twenty volunteers, including not only physiologists such as Dr. J. Negrin, the former Spanish Prime Minister, and Dr. B. M. Matthews, but also a number of working men. Four of our subjects were women. An account is in the press.

The physiological dangers fall under six different heads. The literature concerning (A) and (F), with a full discussion, has been given by Haldane and Priestley.[1]

A. Mechanical Effects

During rapid compression, violent ear-ache, and even rupture of the tympanum, may occur if the pressure on the two sides of the tympanic membrane is not equalized. Most people can easily be taught to do this. Four working men who had never been in compressed air before were compressed to 10 atmospheres (corresponding to 300 feet of sea water) in 5 minutes. A trained subject was compressed to 7 atmospheres in 90 seconds, and this rate could certainly be exceeded. About one subject in five cannot be taught to equalize the pressures rapidly.

During decompression there is less pain, but more danger to life. A number of men have died from rupture of the lungs, which forced air or oxygen into the pulmonary circulation, so that the circulation was blocked by air embolism. This was probably caused by a rapid rise of intra-pulmonary pressure, due to the subjects holding their breath while rising through the water. Any obstruction of the valve by which excess air leaves the escape apparatus would have the same effect. We have had no cases of embolism, but one of our subjects, Mr. J. M. Rendel, developed a pneumothorax.

At 10 atmospheres the density of the air is very striking. The voice becomes nasal, and the increased resistance of the air is obvious even when the hands are moved, and still more so when attempts are made to stir it. The resistance in breathing apparatus

[1] Haldane, J. B. S., and Priestley, J. G., "Respiration" (Oxford, 1935).

may be greatly increased, since the volume of air breathed is unchanged, but the mass increases tenfold, and turbulence may develop, increasing the resistance still further.

B. Nitrogen Intoxication

Behnke, Thomson, and Motley (1935)[1] made the remarkable discovery that nitrogen is a narcotic at high pressures. We confirm their findings. In air at 10 atmospheres all our subjects felt very queer, and many behaved in an irresponsible manner. Manual dexterity was little affected, but arithmetical performance fell seriously in most cases. Some subjects became hilarious; others were greatly alarmed, and thought they were dying. Few could cope with several tasks at a time. There were, however, great individual variations. One subject, H. Spurway, though subjectively affected, was so resistant that her arithmetical performance was actually slightly improved at a pressure corresponding to 250 feet. The symptoms disappear when hydrogen or helium is substituted for nitrogen.

Behnke and Yarbrough (1939)[2] found that argon is rather more narcotic than nitrogen. These results are of importance for the general theory of narcosis; and further experiments with gases such as krypton, xenon, and methane, which are regarded as physiologically indifferent, will be of great interest. It is also likely that at sufficiently high pressures, say, 20 atmospheres or more, hydrogen and helium will become narcotic. These gases would also perhaps reach the threshold concentration for taste or smell, as nitrogen and oxygen do for many people at a partial pressure of about 7 atmospheres.

C. Carbon Dioxide Intoxication

If a compartment of a submarine contained 1 per cent of carbon dioxide, this would not be noticed at atmospheric pressure. If, however, the compartment were flooded at 200 feet, the partial pressure would rise to 7 per cent of an atmosphere. This would make many people unconscious in less than five minutes,

[1] Behnke, A. R., Thomson, R. M., and Motley, E. P., "Psychologic Effects of Breathing Air at Four Atmospheres' Pressure," *Amer. J. Physiol.*, vol. 112, 554 (1935).

[2] Behnke, A. R., and Yarbrough, O. D., "Respiratory Resistance, Oil-water Solubility, and Mental Effects of Argon, compared with Helium and Nitrogen," *Amer. J. Physiol.*, vol. 126, p. 409 (1939).

although fine work, such as gas analysis, is quite practicable in air containing 7 per cent of carbon dioxide at atmospheric pressure. The effects of carbon dioxide and nitrogen are additive. We investigated this question on a number of subjects. Their attitude may be exemplified by the notes made by Dr. H. Kalmus, a Czechoslovak refugee, just before losing consciousness at 10 atmospheres with a partial pressure of 6·5 per cent of carbon dioxide: "This is enough. This is enough.—Not necessarily too much." Consciousness was rapidly regained on decompression, and there were no appreciable after-effects.

D. Oxygen Intoxication

Paul Bert (1878)[1] found that oxygen is a convulsant at high pressures. At 7 atmospheres the convulsion comes on with little warning. There is a slight feeling of anxiety, which would, however, be disregarded under Service conditions. The clonic convulsions are very violent, and in my own case the injury caused by them to my back is still painful after a year. They last for about two minutes and are followed by flaccidity. I wake up into a state of extreme terror in which I may make futile attempts to escape from the steel chamber, whereas, like others, I am quite calm on recovery from carbon-dioxide nitrogen narcosis. Behnke, Johnson, Poppen, and Motley (1935)[2] found that convulsions or syncope developed in men after about forty minutes at 4 atmospheres. We find that all of seven subjects could breathe oxygen for five minutes at 6 atmospheres. At 7 atmospheres, five minutes exposure is about the limit tolerated. It is obvious that convulsions of this sort would be fatal if they occurred while a man was wearing an escape apparatus under water.

E. After-effects of Carbon Dioxide

J. S. Haldane and J. L. Smith (1899)[3] reported vomiting on breathing ordinary air after breathing air containing a high percentage of carbon dioxide for some time. Alexander, Duff,

[1] Bert, Paul, "La Pression Barométrique" (Paris, 1878).

[2] Behnke, A. R., Johnson, F. S., Poppen, J. R., and Motley, E. P., "The Effect of Oxygen on Man at Pressures from One to Four Atmospheres," *Amer. J. Physiol.*, vol. **110**, p. 565 (1935).

[3] Haldane, J. S., and Smith, J. L., "Physiological Effects of Air Vitiated by Respiration," *J. Path. Bact.*, vol. **1**, p. 168 (1899).

Haldane, Ives, and Renton (1939)[1] reported vomiting and severe headache in several subjects after breathing air containing 6–7 per cent of carbon dioxide for an hour or longer. The same symptoms may occur if oxygen is breathed. We have not found such effects after breathing 6–7 per cent of carbon dioxide for so short a period as half-an-hour. Only one of the numerous subjects who lost consciousness when breathing air containing added carbon dioxide at 10 atmospheres even retched appreciably, and this was before losing consciousness, not on recovery.

It is clear that vomiting would be fatal during an attempted escape from a submarine, and it may have accounted for some of the deaths in the *Thetis*. It can be avoided by purifying the air, or by breathing oxygen or pure air for some minutes before attempting escape; this will give time for vomiting to occur if it is going to do so. On the other hand, this danger would not arise after a short exposure to a high partial pressure of carbon dioxide, such as is discussed under heading (C).

F. Bubble Formation during Decompression

This has been the principal physiological danger to divers in the past, and has been fully studied. The tissues take up nitrogen at high pressures. On decompression they become supersaturated and bubbles may form. With very rapid decompression, capillaries in the lungs and brain may be blocked with froth. This causes asphyxia and death unless the subject is at once recompressed. However, such embolism cannot occur if the blood has a reasonable opportunity of unloading its surplus nitrogen. The pressure should never be halved in less than a minute or so, which gives the blood from most organs an opportunity to release its nitrogen in the lungs. With slower rates the main symptoms are "bends," that is to say, pain referred to the joints and bones, and other nervous symptoms such as paralysis and paraesthesia. These are due to the formation of bubbles in the white matter of the central nervous system, and perhaps in the synovial fluid and bone marrow. Nitrogen is a good deal more soluble in lipoids than water, and this may account for the symptoms in question.

J. S. Haldane introduced stage decompression as a prophy-

[1] Alexander, W., Duff, P., Haldane, J. B. S., Ives, G., and Renton, D., "After-Effects of Exposure of Men to Carbon Dioxide," *Lancet*, p. 419 (Aug. 19, 1939).

lactic, and Sir Robert Davis (1935)[1] found that this could be greatly accelerated if oxygen were breathed in the later stages. Even when oxygen is used, decompression lasts for an hour after 15 minutes exposure to 10 atmospheres. Unfortunately, no published figures exist on the limits of safety after very rapid compression to high pressures, followed by rapid decompression, such as occurs during escape from a submarine. We have obtained some data on this important problem. As regards decompression after longer exposures, some of our subjects have had slight symptoms when following the official tables, but these have never been serious. Others can be decompressed much more rapidly without any pain. We do not know the cause of this individual variation. Fatness may be a slight handicap, but I am fairly fat, and have had no trouble when following Sir Robert Davis's schedules of decompression, while thinner men have had "bends" while doing so. Nor do we know the cause of the itching which is almost universal during decompression from high pressures, the rash which sometimes accompanies it, and the rarer symptom of nose bleeding.

Helium has been recommended as a preventive of decompression symptoms, and is used for this purpose in the United States. There is no question that it is of value at high pressures, as it completely does away with nitrogen intoxication. But I am much more doubtful of its value against "bends." Last December I was decompressed according to the Davis schedule after breathing a helium-oxygen mixture at 10 atmospheres. I developed severe pain over a good deal of my body which lasted for an hour or so, and which was followed by itching and "pins and needles" over the area of the skin supplied by the 4th and 5th sacral roots. This was probably due to a bubble of helium in the conus, the tip of my spinal cord. Even after seven months I prefer a cushion to a hard chair, and may perhaps be excused for scepticism of the alleged prophylactic value of helium.

This failure of helium to prevent "bends" throws a good deal of doubt on the current theories as to their causation. Helium is less soluble in water and fat than nitrogen; and whereas nitrogen is more soluble in fat than in water, helium is less so. For this reason it was erroneously concluded that it would be less likely

[1] Davis, R., "Deep Diving and Submarine Operations" (London, 1935).

than nitrogen to produce "bends." The whole problem demands a systematic experimental study with a number of gases. The experiments could be made on animals, whereas experiments on the narcotic effects of gases must be made on men. Animals give very unclear results in this case. Thus a canary flew normally in air at 10 atmospheres, while *Drosophila* refused to do so even when stimulated.

G. COLD

Even when the surface water is fairly warm, the sea may be below 40° F. at a depth of 200 feet. Among the questions which we investigated in this connexion was whether cold increases the narcotic effects of nitrogen and nitrogen + carbon dioxide. Dr. Case lay in a bath of melting ice until, after 12 minutes, he began to shiver violently. He was then compressed to 10 atmospheres, but retained his faculties sufficiently to multiply 47 by 13 in his head. I propounded this question, but was unable to solve it correctly, being more susceptible than he to nitrogen intoxication. A still more drastic experiment showed some adjuvant effect of cold, but it does not seem that any measures need be taken to combat it which would not be justifiable at ordinary pressures.

The main physiological problem to be tackled in planning escape from submarines at depths of 100 feet or more is how to steer, so to say, between the Scylla of nitrogen poisoning and "bends," and the Charybdis of oxygen poisoning. The detailed solution must depend on the details of construction of submarines and escape apparatus, so a full discussion is impossible at the present time. However, it also involves physiological investigations such as those here summarized, some of which will be published in greater detail elsewhere. I am convinced that physiologists have been far too negligent in investigating the limits of human existence, or at least of human consciousness. Physicists often find that mathematicians have already provided them with methods which they need for a theoretical account of their findings. It would be well if physiologists were to investigate the effect of abnormal conditions on human beings before, rather than after, these conditions have killed numerous people, whether in war or in industry.

INDEX